More Advance Praise for
The Wrong Way to Save Your Life

"Reading this book is like listening to stories from a wise, compassionate, and irrepressibly funny friend—one who allows her empathy to fill every unflinching tale about how fear both plagues and saves us. Whether she's writing about gun laws, a bear attack, or postpartum depression, Stielstra's clear voice calls for us to stay awake, and to pay attention."

—Esmé Weijun Wang, author of *The Border of Paradise*

Praise for *Once I Was Cool*

"In *Once I Was Cool*, Megan Stielstra is warm and open and wise. Whether she's writing about the complex loneliness of early motherhood or failing to rise to the occasion or find the right language while living abroad, Stielstra is a masterful essayist. From the first page to the last, she demonstrates a graceful understanding of the power of storytelling. What she's truly offering with her words is the grandest of gifts."

—Roxane Gay, author of *An Untamed State*

"Stielstra's *Once I Was Cool* isn't just edgy, funny, surprising, a ricochet of wow. It's practically actionable. The words reach out from the page. They direct us to look, to think, to ask."

—*Chicago Tribune*

"That is Stielstra's talent: her ability to create experiences. Every narrative seemed to pick itself up off the page and turn itself into a performance before my eyes. The numerous asides, amendments, and annotations force the reader to see and hear her work, not just read it. Ekphrasis (visual description) at its best, there is no contemporary author more vivid in description that Megan Stielstra." —*Chicagoist*

"America's next great essayist is Megan Stielstra." —*Joyland*

Praise for *Everyone Remain Calm*

"Stielstra—collector, curator, and facilitator of so many stories—also writes beautifully and kinetically. Her work possesses a rare aural quality, no doubt the result of so much time on stage, or even in front of a classroom . . . in *Everyone Remain Calm* she gleefully tests the boundaries of the short-story form."

—*Time Out Chicago*

"A daring and inventive debut." —*Collagist*

"Her theatrical performances are intense, composed of a powerful cadence of speech and strong storytelling you won't find anywhere else. Somehow she has bottled the presence of her performances and sprinkled a little bit on each story contained within *Everyone Remain Calm.*"

—CBS Chicago, "Best New Chicago Books"

"[Stielstra] has a profound understanding of where we all go in our minds, and the unique ability to turn it into a story that sounds like your new best friend is telling it to you."

—Elizabeth Crane, author of *We Only Know So Much*

"A trickster constantly unpacking and upending what is meant by 'fiction,' 'truth,' and 'storytelling,' Stielstra has ultimately created a charming style wholly her own."

—Gina Frangello, author of *A Life in Men* and *Slut Lullabies*

THE WRONG WAY TO SAVE YOUR LIFE

ALSO BY MEGAN STIELSTRA

Once I Was Cool
Everyone Remain Calm

THE WRONG WAY TO SAVE YOUR LIFE

ESSAYS

MEGAN STIELSTRA

NEW YORK • LONDON • TORONTO • SYDNEY • NEW DELHI • AUCKLAND

For when you are afraid.

HarperCollins books may be purchased for educational, business, or sales promotional use. For information, please e-mail the Special Markets Department at SPsales@harpercollins.com.

Some of the essays in this book originally appeared, in slightly different forms, in the following publications: "The Wrong Way to Save Your Life," *New York Times*; "Stand Here to Save Lives," *Guernica*; "F," *Midnight Breakfast*; "The Buildup To and Takeaway From," *Soulmate 101 and Other Essays on Love and Sex* and *The Butter*; and excerpts from "twenty" and "forty" in *Great Lakes Review*, *Chicago Artists Resource*, and *Writer, with Kids*.

FIRST EDITION

Designed by Leydiana Rodriguez

Library of Congress Cataloging-in-Publication Data has been applied for.

ISBN 978-0-06-242920-9

17 18 19 20 21 LSC 10 9 8 7 6 5 4 3

Stories can conquer fear, you know. They can make the heart bigger.

—Ben Okri

Contents

CONTENTS

*

I Am Still Fighting with My Big and Small Fears

A lifetime or two ago, I lived with my friends Heather and Pete on Armitage Avenue, just west of Western. Our apartment was a unicorn so far as renting in Chicago: an enormous open loft, wallpapered with windows and cheap as hell. A hallway snaked back to two smaller rooms—one for me, one for Heather—and, behind those, what Realtors call the master bedroom: huge, with high ceilings and a skylight. This was Pete's room. He paid double. He had a nine-to-five job in an art studio with a salary and health insurance, which to me seemed so grown-up and impossible, especially for an artist, but he did it. He made art and he made a living, and if he could, I could. At twenty-three I needed that belief like food and sex and shelter.

When he let me, I'd sit on the floor in his room while he painted, reading aloud from the novels I was studying and sometimes, if I felt brave, my own nervous starts at stories. I loved being there, part of it, the process, the mess, tubes and brushes and sketches and ashtrays overflowing so he'd ash in empty beer bottles or in mugs filled with days-old coffee or mugs filled with paint water or mugs filled with whiskey or wine. I don't think we owned glassware. I don't think we washed dishes. It made more sense to avoid the stacked piles in the kitchen sink and, instead, grab a mug off Pete's floor and scrub out the moldy coffee cream in the bathtub. He didn't have any furniture, just a mattress on the floor, a cage for his iguana, and canvases: finished pieces, half-done stuff—still dripping, newly stretched and ready to go. He painted bodies. He painted abstract. He painted me, circus poster–style, biting the head off a live chicken after I read him the scene from *Geek Love* where Aqua Boy preaches to the devoted: "If they love you then it must mean you're all right. You poor baby. You just want to feel all right."

"Read that part again," he said, pointing at me with a paintbrush. He was muscle and sinew, arms full sleeved with black geometric tattoos. He had a long, pointy goatee a decade before hipsters and what's it called? Beard art. He wore a belligerent uniform of black jeans and no shirt, never a shirt. He must have owned some, at least for work, or our shitty Chicago winters. Band-logo stuff, probably: Slayer and Black Sabbath and Behold! The Living Corpse. I'm trying to see him: reaching down the line of my life, memory as portraiture, a still shot instead of a moving picture. He smelled like cigarettes. He built me bookshelves. He'd make both fists into devil horns, yell "METAL!" and stick his tongue down past his chin. He pointed at me with a paintbrush: "Poor baby. You just want to feel all right."

For my birthday, he took me to the Art Institute. We were there for hours. He taught me to see. I hope that doesn't sound pretentious. It was actually a gift. We went painting by painting: color, context, scale. I'd be ready to move on and he'd tell me to wait, look again, look closer. "What do you see?" he'd say. "How was it made?" "When was it made?" "What was happening in the world when it was made?" "How did what was happening influence what was being made?" "What are you going to make?" and "When are you going to start?"

"Now," I'd say. "Today."

The questions we ask about art are equally vital in examining the self: memory and assumption, love and fear. I'd build on this vocabulary over two decades teaching writers and teaching teachers, but I first started thinking about it with Pete in the Art Institute, standing in front of Egon Schiele, Francis Bacon, Toulouse-Lautrec. My favorite is *Sky Above Clouds IV*, a twenty-four-foot oil painting by Georgia O'Keeffe. It hangs above the back gallery staircase, the ceiling diffused sunlight. It's breathtaking: blue and pink and a sea of white clouds like cobblestones. I see it like a road. A path. You can walk down the stairs or up into the sky.

The truth? I was in love with him.

How could I not be, there under the clouds?

Nothing happened, of course. He was my roommate. He was my friend. He didn't like me like that. It would have gotten weird. I'd been hurt before. Everything ends badly. Breakups are awful. Divorce requires paperwork. Juliet dies in the end. The iguana smelled. I hated metal. What if he didn't feel the same and I had to stop imagining us together, the video on demand as I fell asleep at night? "Uncertainty is better," wrote Chekhov. "At least then there's hope." What if we did get together and I hurt him? I didn't want to hurt him. I loved him. So much so that I'd bug him about the cigarettes and he'd tell me to back off and I'd quote statistics about lung-related death and he'd say we could all go at any time, wiped out by a bus, a train, an explosion like lightning, a heart attack on a mountain, a tumor in the brain, the skin, the breast, by your own hand when it's all too much, or an AR-15 in a school or park or street so who gives a shit about a cigarette, what's healthy or right or fair?

There are so many reasons not to try.

They all start with *I'm scared*.

*

I started writing about fear in 2008. It seems impossible that this period is already history: the Great Recession, a national and global economic decline beginning with the burst of an eight-trillion-dollar housing bubble, dot dot dot.

It sounds like a movie. A novel. Something to read about, not live through.

My husband and I had just gotten married, just bought a condo, just had a baby, the one-two-three punch of the American dream. We were making it: me in education, him in web design. Then—snap your fingers—the market crashed. We couldn't sell. We couldn't rent. We couldn't keep up with the mortgage, which had less to do with the economy and more with me being, quite literally, on the floor. There are so many metaphors about depression in general and postpartum specifically: mountains, climbing over; waves, crashing down; fog, moving through; storms, up up and away. I appreciate the need to find common language, to name the experience and feel less alone in its mess, but I have a hard time seeing the poetry. It sucked, it sucked, it sucked.

Our condo was across the street from the Aragon Ballroom, a legendary rock club in Uptown on Chicago's North Side. Bands played every weekend. Fans lined up around the block.

You could hear the music through my windows, a weird sort of soundtrack to the most scared I've ever been. Marilyn Manson and my baby won't eat. Smashing Pumpkins and my baby won't sleep. Yeah Yeah Yeahs and the banks call every day: Do you know you're overdue? Do you know the consequences of being overdue? Are you sure? I was scared to leave the house, scared I would hurt myself. *Something is wrong, I can feel it, no it's not all in my head.*

At one point, I opened my journal and wrote: I NEED HELP.

*

I climbed the stairs to the apartment on Armitage and covered my ears. The music from Pete's room was so loud the apartment vibrated; a car with the bass cranked, a strobe light with sound. In my memory it was Rammstein. German industrial metal. They play a subgenre of rock called neue deutsche Härte or new German hardness; their name translates to "ramming stone." That's what my bathroom sounded like: hardness, ramming, a jackhammer.

I went into the bathroom where Heather was getting ready: mascara and lipstick and wild red hair. She worked late-night bar shifts; I worked daytime brunch. We saw each other at 3:00 a.m. if I was up writing or 3:00 p.m. before she ran out the door. We watched *Ab Fab* and wuxia films. She took me to buy my first vibrator and my first window air conditioner, necessary appliances for the modern woman. I'm trying to see her: reaching down the line of my life. She smelled like honey. She loaned me dresses. She gave me music: hip-hop, house, and R&B, which in 1998 meant *The Miseducation of Lauryn Hill* and in Chicago meant Jesse De La Pena. We'd go dancing at the Subterranean and I'd study her: hip and swerve and that wild hair. I tried to move like she moved—hands go here, feet here, ass like this—the same way I learned dialogue from Hurston, pacing from Selby, sentences from Woolf. Heather wasn't having it. She tried to help, to loosen me up, putting her hands over my eyes and instructing me to feel. "Listen," she said, her lips at my ear. "Being in the music is like being in a book." I've danced with that idea for years: feeling as reading, the body as a text.

"WHAT THE HELL?" I yelled, gesturing at the noise.

"I KNOW," Heather yelled. "SOMETHINGSOMETHINGSOMETHINGLOUD."

"WHAT?"

"I TRIED TALKING TO HIM BUT—"

"WHAT?"

"HE TOLD ME—" she started, then gave up and made follow-me hands. We went down three flights of stairs to the driveway and leaned against her minivan, a peach-colored thing she lent to me for teaching gigs. We could hear the music through the windows but from outside it was muted, pulse instead of punch. A man sold fresh melons from a pickup truck out front. Kids chased an elotes cart down Armitage. Heather lit a cigarette and I reached for it; I didn't smoke but what else do you do with your hands? She sounded worried as she told me a friend of Pete's had died. She didn't know who. Someone from before us. Someone he loved. "He went into his room and turned on the music," she said, taking back the cigarette. "I tried to talk to him, but—" she stopped because truly, what do you say? We hold our friends' hands as their hearts break: lost lovers, lost children, divorce and illness and addiction. There are no perfect words. We can be there. "How can I help?" we say. We say, "I'm sorry."

"Is he okay?" I asked.

We both looked up, third floor, back of the building.

What a stupid fucking question.

I don't remember how long it went on: the jackhammer, the locked bedroom door. I want to say weeks, but I'm known to exaggerate, to add the extra ten feet the hero has to jump. I'd come home from night classes, from student teaching, from waiting tables, from writing in coffee shops, from hanging out at friends' houses so mercifully still and quiet and every time I climbed the stairs, I felt the floor beneath my feet.

"Pete!" I'd yell through his door and over the music. "Can I get you anything?" and "Do you want to talk?" and "The music is really loud!" But what I meant was: *How can I help?*

"No," and "No," and "I know," he'd yell back. But what he meant was: *Leave me the hell alone.*

One night I came home and heard nothing. No jackhammer, no pounding bass. At the time, I was reading the novel *So Far from God* by Ana Castillo. I was at the part where Fe—the middle sister who'd been screaming for months, stopping only to sleep, then waking up and screaming again, "her bloodcurdling wail became part of the household's routine so that the animals didn't even jump or howl no more"—had suddenly stopped. There was silence. So loud a silence. I heard that book in my own apartment. My life collided with its pages.

The shower turned on in the bathroom and, at the end of our snake-like hall, I saw that Pete's door was open. I tiptoed—past my room, past Heather's, and into his.

Twenty years later, and this is what I see:

The room wallpapered in dull brown, what looks like grocery bags meticulously cut at the seams, then laid flat and taped together. Look closer and there's the imprint of folded rectangles and the thick, clear industrial tape. Over that is the paint. Dark, dark purple. Dark red and blue. Thick and globbed, like it was thrown at the wall or sludged on with a knife. It starts at the ceiling and sprays down, running off the paper and onto a tarp over the floorboards, less like you're looking at it and more like you're inside it. I wanted to get out of there. But at the same time, I couldn't look away: a car crash, a bar fight, a shooting, a scandal. We have to watch. We have to see.

"It's a lot. I know." Pete stood in the doorway, a towel around his waist and dripping from the shower.

I said something then. I don't remember what. I hope it was: *I'm sorry.*

We both looked at the walls.

"I wanted to get it out of me," he said, and even though it would be years before I understood what he meant, before I put my own heart in the open and looked, *really* looked, I still nodded like I understood. "You can't fix it if you can't see it."

*

I don't know how history will remember the summer of 2016, but it will be impossible to talk about it without talking about fear.

I want to place you here. Something happens—say, Britain votes to leave the EU or Beyoncé releases a new album or a candidate for the United States presidency talks about his dick on live television. And within the week, hell, by the end of the day, there are op-eds and essays and—what are we calling journalism now?—content. Usually I'm writing in rapid response, so it's strange trying to describe the culture of fear leading up to the 2016 election knowing that by the time you read this, the decision has already been made. Hillary Clinton may have served her first hundred days. Donald Trump may be signing an arms agreement with Russia. The GOP may have split and now we're a three-party system—maybe four parties, maybe five. At this stage in the game, nothing would surprise me. Secession. Aliens. Skynet. We're at the part in the sci-fi movie when machines take over the world because humans are killing each each other and they want to save us from ourselves.

It would be fascinating if it weren't so terrifying.

Twenty-four-hour news cycles and scrolling social media feeds updated to the second, no time to process, to breathe. In Orlando, forty-nine queer people of color were murdered while they were dancing. Residents of Flint, Michigan, are still drinking lead-contaminated water. So far this year, 1,023[*] people have been shot by the police. At Trump rallies, the predominately white crowds yell awful slurs at black, Latino, Muslim, and gay people, and shout in reference to Hillary Clinton: "Hang the bitch." A ten-year-old child yells, "Take the bitch down." In August, the Southern Poverty Law Center

* "The Counted," *The Guardian*, accessed July 5, 2016, https://www.theguardian.com/us-news/series/counted-us-police-killings

issues a report documenting how Trump's violent rhetoric is poisoning our nation's elementary, middle, and high schools, "producing an alarming level of fear among children of color and inflaming racial and ethnic tensions in the classroom." On MSNBC, Cokie Roberts takes the question to Trump himself, bringing up incidents of white children telling their darker-skinned classmates they'll be deported if he wins. "Are you proud?" she asks. "Is that something you've done in American political and social discourse that you're proud of?" He tells her it's a nasty question.

I want to believe we're moving forward as a society and this surge of bigotry and violence is the final kick and scream, a last-ditch effort to hang onto the white patriarchal systems that favor the privileged.

If we're going to make it, we have to look at the fear.

We have to get into it. Throw it against the wall, stand back and take a good close look. It's ugly: heavy, dark, and centuries in the making. You might want to move on, to turn it off, watch something else, but wait—look again. Look closer. How was it made? When was it made? What was happening when it was made? What are you going to do about it? And when are you going to start?

Now, I think.

Today.

ten, or The Little Girl Character

3

The first thing I remember is fear. I'm in the air—can't move, can't cry, breath locked. All I hear is heartbeat. I look down and there's my dad reaching for me, his face blurry below his open hands.

Later, my mom filled in the blanks: This was Alma, Michigan, the town where I was born. Our neighbors had a treehouse, one of those epic little-kid fantasylands of ladders and rope bridges—think *Swiss Family Robinson* or Finca Bellavista. I was always getting stuck, going up and afraid to come down. I don't remember any of it—not the tree, the neighbors, or even myself as the little girl character in this childhood narrative, typical small-town white and mid-middle class. All that's there is the fear—my breath, my body, my bones.

"Jump, kid," said my dad.

I'm forty years old and I can still hear his voice.

"Jump."

5

New town, new house: forty-some miles south to Owosso. In the side yard was a creek full of thorns and wild blackberries, my fingers dyed forever purple with juice and blood. Fairies lived in that creek. I believed it then and I believe it still. Once, I caught a bucketful of frogs and kissed them one by one. They were bumpy and cracked dry. I remember thinking they needed ChapStick. I remember wondering why none of them turned into a prince. Was something wrong with my kisses? Was I the wrong kind of girl? We don't question the fairy tale, we question ourselves, and I was mad at Tinker Bell and Cinderella and their dumb, frothy tulle. Then my mom read me the original Grimm's. My favorite part, then and still: when they cut off the stepsisters' toes to fit into that goddamn glass slipper.

Between the creek and our house was a hill. I recently drove past and was surprised to find it so molehill small compared to the Everest in my memory: impossibly steep, miles high in the clouds. This was the scene of my recurring nightmare, not the first-person kind where you're in the dream, but the third, where you watch yourself: a little girl in OshKosh B'Gosh and a yellow bowl cut, running down that enormous hill. She's so scared. She stumbles, hits the ground, and scrambles up looking over her shoulder. Behind her, chasing her, close, closer, furious and growling and flexing his ridiculous green muscles is the Incredible Hulk, specifically Lou Ferrigno from the 1978 show on CBS. People laugh when I talk about this, but truly I was terrified. I can still hear it: the frenetic piano from the opening theme, the sound of fabric splitting as Banner morphed into the hulk,

Ted Cassidy's deep, scary/sexy voice-over: "Until he can control the raging spirit that dwells within him." I was scared to let a foot dangle over the side of my bed; he'd grab it and pull me under. I was scared of the basement; that's where he lived, lurking behind the washing machine or the piano or the tiny black-and-white television, its rabbit-ear antennae wrapped in foil. I was scared to fall asleep, to enter that dreamworld with the running, the stumbling, so close, closer, muscles popping green and veiny. As soon as I got to the bottom of the hill, it would all start over at the top.

For years, I carried that dream. I can still see it, a video on demand that as the years bled by was replaced with more garden-variety Psych 101.

I'm about to go onstage. I don't know my lines. My script is missing pages. "That's your cue!" someone says, and I'm pushed into the spotlight.

I'm running around the restaurant, arms stacked with plates. I've forgotten the eight top at table twenty-four. When did they seat table twenty-four? Why didn't anyone tell me? Everyone wants red pepper benedicts but we're out of red pepper benedicts so customers are pissed, standing and yelling, their children running around unsupervised. One of them knocks into me. I drop a pot of boiling coffee.

I'm standing in front of hundreds of students and I can't speak. My mouth is full. I reach in and pull out wads of gum, long and stretchy like taffy. I pull and pull and pull, an enormous pink pile growing at my side. Everyone sits, watching, waiting for me to say something, to quote somebody, to talk, to teach, but I am choking, both hands at my throat and I wake up fast, the movie cliché of sitting up in bed, sweat coated and gasping.

It's been forever since I remembered my dreams.

Who has time to sleep?

6

I was dancing with a boy and his dog attacked me. It was a dachshund, tiny and mean. I don't remember its name. I don't remember the boy's name, either. His parents were Gloria and Lenny and I believe their last name was Kravitz, but maybe I'm thinking Lenny Kravitz the musician Lenny Kravitz. They were kind. They laughed a lot. They were the first people I knew who were Jewish. The only grown-ups I ever called by their first names. Lenny was a private investigator—I think that's right. There were a lot of families on the block. I could be mixing things up.

This much is true: the boy had cool stuff,* specifically a plastic Fisher-Price record player. Do you know the one? Red, with a yellow arm? It's retro now. On Walmart's website they say it "references the funky playthings of the past." We were dancing to some album, that little-kid frenzy of jump-

* He also had every Masters of the Universe action figure including Rattlor and Two Bad and Zoar the fighting falcon, however—and this is very important—he did not have Castle Grayskull. So one time he and his friends told me I couldn't play with them because I was a girl and Masters of the Universe were for boys and I went home crying and told my parents, and even though money was tight, even though they weren't fans of Barbie-type plastic toys, even though they believed in solving problems with logic and discussion ("Use your head," they'd say, "your words"), we went straight to Kmart and bought Teela and Evil-Lyn and the Sorceress. (*Where are their pants?* I wondered. *Aren't they cold?*) And—the icing on my six-year-old fuck you cake—that ginormous plastic castle.

I was hot shit on the block, I tell you what.

Know what else? I shared.

ing and thrashing. His dog must've thought he was in danger. Its muzzle curled, the growl and snarl before it jumped. The bottom teeth caught me under the chin, the tops just below my ear. I don't remember blood and I don't remember pain, although the look on my mother's face made it clear there was plenty of both. This was an ongoing theme of my childhood, and—let's be honest—to this day, when I'm supposed to be so grown-up and independent. I see my mom and instinctively match her emotional response, be it fear or sorrow or joy. I'm hyperaware of this now, my own kid six and seven and eight years old: he'll crash-land, register the damage—skinned knee, goose egg?—and look at me, not for help, but to figure out how he feels. If I panic, he panics. If I'm scared, he's scared. If I'm like: "Hey! No big deal! Get back on the skateboard!" he's back on the skateboard. I imagine parents of twelve-and thirteen-and fourteen-year-olds saying, "Enjoy it while it lasts." But I'm thirty-eight and thirty-nine and forty and still, when my mother is scared, I'm scared; when she's happy, I'm happy; when she cries at the mere mention of that Hallmark commercial where the big brother comes home from war on Christmas morning and sings "O Holy Night" with the little brother, I'm a wreck, too. I'm wrecked now, crying at my laptop in a coffee shop in Evanston, Illinois. Whose tears are these—hers or mine?—this mother bond tying us together across the miles, the years.

We are both, to this day, uneasy around weenie dogs. I'm sorry if you have a weenie dog. I'm sure your weenie dog is great. I have a pit bull, and I get pissed as hell when the entire breed is blamed for the actions of one. Believe me, I'm aware of my hypocrisy. Intellectually, I'm not afraid, but the mind's got nothing on the gut. I see one of those yippie little battery-operated hot dogs and I'm out of my seat, backing away, the thump-kick of heartbeat audible across the miles, the years.

6

I was an only child, which meant I was lonely but also that I had magical powers. I could talk to people inside the television. I could whisper to the sky and start a thunderstorm. I could move stuff with my brain because I was actually a Jedi. I begged my mom to take me to movies about Leia, ones where *she* got trained in the Force and *her* light saber shot into her outstretched hand and *shwwwooop* lit up blue and electric and ready to save us all. I cried my eyes out watching *The Force Awakens*. I'd been waiting for Rey since I was six years old.

But above all else: I could bring back the dead.

We kept rabbits in a hutch behind the garage and one of them got pregnant in the winter. My dad built an insulated nest, hay stuffed and heated from below with a lightbulb for a furnace, but the mama bunny gave birth just outside of it. When I opened the hutch the next morning there was a pile of dead bunny popsicles, ice crystals like freezer burn on the thin pink skin. I immediately forgot everything my father had taught me hunting pheasants, the conversations my mother and I had after reading *The Big Wave*, the leaves turning colors and drying up in the fall, drawing maps of the Mesozoic era in kindergarten class, life and death intertwined in a big old circle like a snake eating its tail and I looked at those tiny bodies and thought, *Get up*.

I said it out loud: "Get up."

I waited. I waited hours. I waited days.

Part of me is waiting still.

6

I was very sure I would die playing Don't Touch the Floor, that game where you pretend the ground is red-hot lava.

7

My dad studied Henry David Thoreau, who wrote that "one achieves an ideal spiritual state via personal intuition rather than religious doctrine." My mother, an early childhood educator, told me we'd examine many different religions all of which had a rich cultural history that would teach me so very much about our wide, wild world and later, when I was a grown-up, I could choose whichever I wanted.

No way could I wait to grow up.

"What happens at your church?" I asked a girl in my first-grade class, some princess in kneesocks who inexplicably kept her whites white.

"Lots of things!" She was sitting on the reading carpet brushing the tail on a My Little Pony. "People sing about Jesus and line up to eat wafers and have religious experiences."

"What's a religious experience?" I asked. I was fascinated, partly by the religion but mostly with the pony.

"It's when Jesus speaks inside your brain and you know he's with you all the time but not in a creepy way," she said. The pony had glitter on its butt. Its tail was blue and shiny.

"And also when you say 'he' the *h* is capitalized. That's really important." She looked at me, suddenly worried: "Don't you talk to Jesus?"

I shook my head no.

"I'm sorry," she said, her mouth an upside-down U. "You're going to burn in hell."

That night, before bed/after bath, I went to the line of green leather-bound encyclopedias on the bookshelf and counted volumes up the alphabet: Conifer–Ear Diseases, Earth–Everglades, Evidence–Georgian S.S.R. I pulled down Geraniales–Hume and sat crisscross applesauce on the oriental rug, then flipped through pages like my mother showed me: G to H, H–A, H–E; heelwalker (insect); Hefner, Hugh; Hekla the Icelandic volcano; H-E-double-L hell. The context was over my head, of course, but there were lots of words I recognized—darkness, dead, shadows, demons—and pictures both terrible and fascinating: Signorelli's *The Condemned in Hell* from the chapel of Saint Brizio and the right-hand panel of *The Garden of Earthly Delights*, both depicting agony and pain and things generally traumatizing for seven-year-olds and grown-ups alike.

"Yes it is true!" I told my mother, fully freaked out. "It's in the encyclopedia!"

This is a memory I go to when I consider the type of parent I'd like to be. She sat me down and explained that she'd left Catholicism at eighteen because she didn't believe in hell. She showed me other books, other pictures, other ideas of what the afterlife could be. And she gave me an age-appropriate sort of lecture that boiled down to: check your sources.

7

Fire drills were easy: get up, go outside.

Tornado drills: those were more complicated.

The emergency siren starts screaming. Teachers tell everybody to remain calm, this is just a drill, and your first-grade class stands up in an orderly fashion and lines up at the door. Classes from upstairs are coming downstairs because downstairs is safer so the hallways are packed, students like sardines with our backs against the lockers. Stay away from gyms and auditoriums. Stay away from glass and windows. Slide down to the floor, tuck your head into your knees, and put your hands over your neck. This is to prevent paralysis in case anything falls you are told. When you ask what will be falling, you are told debris. When you ask what debris means, you are told: *Shhhh be quiet. Pretend this is a real tornado.*

I was very good at pretend.

I saw funnels of whipping wind slam into our school, ripping off roofs and flinging children to their deaths. I saw jagged shards of broken window flying through the halls, people curled at the waist to protect their internal organs, grabbing hold of each other to keep from being swept away in the furious wind. I saw office supplies like airborne missiles, my friends flattened under upturned furniture, my parents searching through our decimated building to find my tiny, broken body in the rubble, and I sobbed into my knees, locked in a ball with my back against the still-there wall.

Now our kids have active shooter drills.

I wonder what they pretend.

7

I didn't see my dad much during the week. He worked as an elementary principal in Perry while getting his associate's in Lansing; later, he stayed in Michigan while my mom taught in Colorado; later still he commuted between my school in Owosso and a new job in Chelsea, so weekends were our thing. We went hunting, mid-Michigan forests and endless dead-grass fields. We went fishing, eleven thousand inland lakes with walleye and bass and trout. We went canoeing on the Shiawassee River—me in the front, him in the back, dog in the middle. Once, we went over a waterfall and the boat flipped. It was thrilling, my first remembered adrenaline. I came up for air and saw my dad swimming toward the dog. He didn't have to take care of me because he knew I was a good swimmer. He'd taught me, throwing me off piers and into deep ends before I could walk. He trusted me, believed in me, and if he believed in me, I could believe in myself! I could kick through the frothy water, get myself to shore, and what a great metaphor for life!

Recently, I recounted this memory to my dad—the waterfall, the swimming, the metaphor—and he looked at me like what the goddamn hell. Of course he swam to grab me when the boat flipped! We were in the rapids! I was just a kid! Zeke wasn't in the water. He hadn't even been in the boat! We'd pulled to the shore to let him out and then we went over the waterfall—because you always go over the waterfall, that's what a waterfall is for—and he grabbed me underwater, got me to land, and went back for the canoe, hustling 'cause my snowsuit was wet and he was worried I'd—

"Snowsuit?" I said.

"It was November," he said.

"We went over a waterfall, in Michigan, in November?"

That's what a waterfall is for.

"You were freezing," my mother said when I brought this up to her—the rapids, the canoe, the snowsuit. "You were purple. I thought you had frostbite. I thought we'd have to go to the hospital. I got you out of your clothes and put you in the bathtub, cool water first till your body adapted to the temperature and—"

Their stories tangle together, the he said/she said of my childhood, but what happened next is a memory all my own. My dad came into the bathroom and my mom pulled the shower curtain, as though that thin layer of vinyl would shield me. Then she lit into him, loud and furious. I sat naked under the tepid faucet, learning all sorts of things like: grown-ups make mistakes; grown-ups get mad; grown-ups yell.

Their divorce wouldn't happen for a decade, but still: this is the only time I remember hearing them fight.

7

My son is seven. It's no longer possible to have a conversation over his head. He hears us talking and wants to know—"Mom, what's a primary? Who is Laquan McDonald? Where is Syria? Why can't our friends go the bathroom? What's a trump?"—and I take a deep breath and do my best.

When I was seven, the question was: "What's a pink slip?"

Michigan was cutting public school teachers like crazy. My mother couldn't find a job with my dad's school in Perry, or near my dad's school in Perry, or even in the same state, so they both applied for everything listed on the MSU job bulletins. A position came up in Denver, fourth grade with a focus on integrated learning, and the headmaster wrote my mom and asked if he could fly her down for an interview.

"I wasn't sure what a headmaster was," she told forty-year-old me. "But I had seen *Goodbye, Mr. Chips*, so what the heck."

"Wait until you see the library," she told seven-year-old me. "It has a hundred thousand books!"

The plan was this: Mom and I would go to Colorado for the academic year while Dad stayed in Michigan, both of them applying wherever they could. You see this all the time in education, both K–12 and the academy: one partner gets an offer, the other doesn't, and what happens next? The relationship ends, the family splits, a career is sidelined, or you do the distance with your fingers crossed. Something will come up, right? Right?

My memory of this time is vague in detail but clear in emotion: It sucked. I missed my dad. I had awful nightmares. The city was always on smog alert. Money was tight tight tight. Since my mom worked at the school with the headmaster, I got tuition remission, but we both had trouble with culture shock: kids getting dropped off in Rolls-Royces, parents who were dignitaries and gone months at a time, my mom doing parent/teacher conferences with nannies, second graders with second and third houses in foreign countries called Nantucket and Napa and the Hamptons. One time, I was invited home after school with a girl in my class and she made me eat a bar of soap. We were in her bedroom; it was on the fifth floor. She told me to go into the bathroom; she had her own bathroom. I should get in the shower; she had her own shower—with glass walls like Willy Wonka's elevator. "Pick it up," she instructed, pointing at the soap. "Pick it up and eat it." She stood there, watching me through the glass. I remember my front teeth sinking into the hard fat—green and scented sickening, a rich man's Irish Spring. The texture was nails on chalkboard. It's still in my mouth, still choking down my throat, chunk and bile and chemical detergent, but I kept going, staring back at her and chewing slow.

I didn't eat the whole bar. Half, maybe. Then: "You may stop," she said, and went back to her bedroom to play.

My mother cried when I told her, so I cried, too. She must have asked why I did it. God, what did I say? Because I was the guest? Because I thought this girl was better than me? Because I was scared? I don't remember. What I do remember is the look on my mom's face. I knew we'd be getting the fuck out of Denver.

I didn't go to anybody's house after school after that. I waited for my mom in the library. It really was incredible.

8

The whole year I was in Denver, my life in Michigan stayed the same: our house, scary basement, rabbit hutch, Everest, and my best friend, Sara, without the *h*. She lived on the other side of the creek four houses down, and I was allowed to walk there by myself. Her parents treated me like one of the family and ohmygod they were tall. I had to bend my head all the way back. They had a swimming pool and my hair was always green from the chlorine, sometimes mermaid, sometimes alien. In their living room was an L-shaped couch—L-shaped, it blew my mind!—and, get this, MTV. We'd sit cross-legged, jaws on the floor, watching the video for "Billie Jean" by Michael Jackson. We were too young to get the lyrics and too small-town white to understand the magnitude of the first black artist in prime-time heavy rotation. All we heard was his voice. That three-note synth. The pink shirt, the black-and-white loafers. When I walked home after dark, four houses back and across the creek, I'd imagine the ground lighting up beneath my feet. Instead of being scared of what was hiding in the bushes, I danced down the asphalt.

9

My mother has given me many gifts. The best? A library card.

She let me pick out anything, everything. I'd show her the books, and she'd read them, too, and we'd talk about them: what I learned, what confused me, what was new enough to sound strange. I didn't know some kids didn't have parents until *The Great Gilly Hopkins.* I didn't know girls weren't supposed to go hunting until *Island of the Blue Dolphins.* "Don't be afraid to be afraid," I'd tell myself like Meg Murry in *A Wrinkle in Time.* I remember crying my face off when Aslan died in *Narnia*, and the dogs in *Where the Red Fern Grows*, and Leslie in *Bridge to Terabithia.* "It's okay if a story makes you sad," my mom told me. "It's okay if it makes you angry or afraid. Those feelings are real. Let's live them."

After we checked out our books, we'd go next door to Wendy's for Frostys. I wasn't typically allowed stuff like that, junk food or fast food or pop, and those Frostys tasted dangerous. Forbidden. I didn't do much that was forbidden, didn't sneak around and get in trouble and test boundaries as kids so often do. For better or worse, I stayed in my head.

The world outside seemed less shiny—dull and on mute.

A lifetime later—after we'd moved away, when I was a

teenager with a bra and St. Ives Blemish Control Peach Scrub and Girbaud jeans rolled three times and pegged so tight at the ankle it's a wonder we still have feet and whatever book I was reading at the time, my favorite being *Dune,* a sci-fi novel by Frank Herbert about interstellar feudal society with a powerful order of women called the Bene Gesserit who trained their minds and bodies to the point of superhuman powers, which was so totally cool—my parents and I drove through Owosso and stopped at that same Wendy's. I took out my retainer and wrapped it in a napkin. I remembered it two hours later, in the back of the car on the way to who knows where. Dad turned the car around and back to Owosso. Orthodontia is expensive and rarely covered by insurance. We spent hours in the alley behind the restaurant, searching through trash bins of discarded Wendy's. We smelled like Wendy's. We had Wendy's in our hair. My mother tells me that we found the retainer eventually, but all I can remember is the fear of disappointing my parents. To protect myself against their iron silence, I repeated the Litany against Fear, used by the Bene Gesserit to focus their minds during great peril.

I must not fear.
Fear is the mind-killer.
Fear is the little death that brings total obliteration.
I will face my fear.
I will permit it to pass over me and through me.

9

New town, new house: fifty-some miles south to Chelsea, which, as of 2014, is officially a city but back then was most certainly a village. Home of Jiffy mix, the Purple Rose Theatre Company, and the greatest public library in the history of the universe. I loved that library. I loved Thompson's Pizzeria. I loved the flag girls that danced in front of the marching band and the demolition derby at the Chelsea Community Fair and the fact that both my parents had jobs in the same place; Mom as the gifted education specialist, Dad as the middle school principal. On weekends, he'd have to work, and I'd run through the empty hallways and the empty classrooms and the empty gym.

Our new house was a ten-or fifteen-minute drive from downtown on one of the many small lakes in the area. There were a handful of residential properties, but mostly it was state land, some cleared to rent for summer camps and retreats, some running wild with mossy, swamp-like forestland full of poison ivy. I learned where to walk and where not to walk. Outside, our house was nondescript, white with blue shutters and hidden in the trees; inside, it was like a log cabin, wood walls and floors and ceiling with an unfinished attic for my bedroom. My dad built me a window seat. My mom let me paint poems on the walls. There were built-in bookshelves and a closet for record albums, Beethoven and bluegrass. The garage was its own world: fishing poles and tools hung from pegboards, a second refrigerator packed with frozen meat and Ziploc bags of blueberries, canoes and toboggans in the rafters, stacks of lumber and a standing saw, rifles in locked cabinets, boots and camo and rubber overalls, box frames to house beehives in the summer, and mounted heads, mostly white-tailed deer. I named the biggest one Bilbo Baggins. Sometimes I'd sit out there and talk to him.

Do you think that's weird? Remember, I was an only child.

My parents both worked. There weren't any kids my age around and I was still too young to bike downtown alone. It was easier in the summer. I had the lake, with a raft in the middle on floaters. As you swam toward it, seaweed reached up from the bottom and tickled you, which felt fine at first but later, at somebody's slumber party, we watched a grown-up horror movie where a woman went swimming at night and the seaweed came alive, winding round her legs and arms and pulling her screaming into the black. I didn't swim to the raft after that.

Ever.

Still.

I take the goddamn rowboat.

9

I walked into class, looked around at the new, strange kids in their rows of desks, looked up at the teacher, said, "Mr. Bullock?"—and puked. I puked everywhere. Over everything. This is not an attempt at dramatic effect, an intentional use of exaggeration to make a better story. No, there was puke on the blackboard, on the bulletin boards decorated with construction paper. Puke got on kids. It got on their desks and they jumped and screamed. Mr. Bullock walked me down the hall to my mother, who worked at my same elementary school. She helped me clean up and get back to class, but it was too late. I was ruined. For weeks, no one would sit next to me. When I raised my hand, they'd duck and cover. When I opened my mouth to speak, they'd make puke sounds like Chunk in *The Goonies.*

But it was nothing compared to the multiple sclerosis read-a-thon.

For one week, once a year, we all wore stickers that said KICK MS, and for one week, once a year, I got the shit kicked out of me. When I tell that story, people laugh. Probably because of how I tell it—humor as weapon, humor as shield. But here's the truth: gingerly pulling down my pants so the elastic from my underpants wouldn't snap against the blue-black welts across my hips; sitting sideways in an elementary-kid-size desk 'cause straight down on my butt hurt too bad; walking through the hall and staring straight ahead, ignore it, keep going, pass over me and through me. But this is true, too: I was lucky. For me, it would be over in a week. It was so surface, so stupid—my initials, for god sakes. But for other children, this happens every day because of the way they look or talk or dress, where they live or where they're from or who their parents are or any of a number of reasons why we're all so impossibly cruel.

10

My dad asked me not to tell anyone about this, but he's retired now, hasn't worked in a school for a decade, hasn't worked at that school for two decades, and it's been three decades since it happened so the statute of limitations is up.

One night, he used his master key and took me swimming at the Charles S. Cameron Pool.

It's big: a six-lane, twenty-five-yard competitive pool with a separate shallow end for swim lessons. Underwater lights illuminated the blue floor of the pool, but we kept the main lights off in case the janitors were still in the building. At the time, there were three diving boards: two standard size and, between them, the high dive. You've seen this moment in the movies: The girl stands at the base of the ladder, nervously tugging at the butt of her Speedo. She looks up, up, up at the terrifying height, trying not to let on how scared she is. This fear will follow her for the rest of her life: airplanes, mountains, roller coasters, the Skydeck at Sears Tower. At the top, she clings to the poles at the side of the board and scoots forward till they're gone, just her and the darkness and the low-lit water a million miles below. The board trembles beneath her feet. All she hears is heartbeat. All that's there is fear—breath, body, bones.

"Jump, kid," said my dad.

I'm forty years old and I can still hear his voice, bouncing on the water like an echo chamber.

"Jump."

Here Is My Heart

Write your name here. Address, here. Here—check every box on this long list of disorders and diseases and conditions that are a part of your medical history, your parents' medical history, your grandparents' medical history, and down the DNA. So much terrifying possibility. So much *what if* in our blood, our bones.

I checked two. Melanoma and—

"Heart disease?" my new doctor asked. I liked her immediately: her silver hair, her enviable shoes. Later, I'd love her intelligence and, later still, her respect for my intelligence even when—especially when—I acted bonkers. She removed the weird, spotty growths from my arm and told me they weren't cancer. She diagnosed my thyroid disorder and fought it like a dragon. She helped me understand my own body and demanded that I treat it with kindness, even when—especially when—I was stressed or exhausted or scared. It's so easy to forget ourselves, to prioritize our own hearts second or tenth or not at all. Do you see yourself in that sentence? Are you, right this very moment, treating yourself less than? *Cut that shit out*, my doctor would say, except she'd say it in professional, even elegant doctorspeak. And to her, I listen. Her, I trust. Every woman should have such an advocate and the fact that our patient/doctor relationship is a privilege as opposed to a right makes me want to set the walls on fire. Look up—see the wall in front of you? Imagine it in flames.

"Megan?" she said, and I pulled myself away from her shoes. "There's a history of heart disease in your family?"

"Yes," I said. "My dad."

She asked questions and I did my best. "In his forties, he had chest pains shoveling the driveway in Michigan and got—are they called stents?—he got stents, and then in his fifties he had the same pain playing racquetball and they had to medevac—"

"Medevac?"

"This was in Alaska—he lives on an island now. It's called Kodiak. Yes, there are bears. They sent him to the mainland for surgery and afterward he went right back up the mountain—"

"The mountain?"

"He's always on mountains, hiking insane heights with drop-offs like this"—I made a ninety-degree angle with my hands—"and carrying moose around in backpacks that are like what, a thousand pounds? And sure, fine, he's in great shape; he eats tons of wild salmon and can bench press you and me put together but still, he's almost seventy, and a week ago he was hunting and had those pains, again, and they sent him to the mainland, again, and—"

The more I explained, the angrier I got.

Anger is easier than fear.

Afterward, I called my dad's cell phone. We talk often, a couple times a week. He tells me about the weather or the ocean or whatever book he's reading, Kate Atkinson's *Life*

After Life or Ron Chernow's biography of Alexander Hamilton, but that day I wanted to hear how he was doing. He just had heart surgery. I pictured him on the couch, dog at his feet, benched for who knows how long, and bored out of his mind. Maybe I could entertain him, tell him stories, make the time pass quicker. Healing sucks.

He picked up on the first ring.

"Hi! How are—" I started, but I didn't make it to the you.

"Can I call you back?" he said in an exaggerated whisper. And then, in the same matter-of-fact way you or I might say I'm at the store or sitting down to dinner, he said, excited, "I'm tracking a moose!"

*

The first time I dissected a heart was a miserable failure. Me, in my kitchen in Chicago with a knife and a fist of raw meat, and then, a minute later, a long, veiny slab, veal-colored and flat like skirt steak, watery purple seeped into the cutting board.

I remembered the frogs I dissected in high school, the smell and the death and the tiny, fragile jawbone; deer carcasses hung upside down and bleeding in our garage in Michigan; Christopher, my then boyfriend/now husband, gutting rabbits on the porch to impress my father, the blood and the mess and the pressure of becoming a family; laying on the table in my paper robe and hearing my son's heartbeat: *He's alive he's alive he's alive.*

So much joy. So much fear. How can a heart take it?

I stared at my work, messy wet chunks of meat. After a while my dog stuck his nose in the back of my knee. I looked down and he wagged his tail so hard his shoulders shook.

"Fine," I told him. "You win," and I set the cutting board on the floor.

*

Every fall, my dad flies through Chicago to hunt birds in Michigan—him, my uncle Chuck, and two English setters to point the way. I love these trips because we get to see him and I hate these trips because we don't get to see him enough: morning flying in, evening flying out, and a few days between hunts. Poor Dad. Poor Chuck. I've yelled at them both.

Maybe I used the wrong words. Let me try again.

I love you. I miss you. Stay another day. Stay two days, a week. Your grandson is getting older. I'm getting older. You're getting older. I have so many questions, so many things still to learn, so many things to say like: *Please, Dad, please, I'm afraid of the mountain*. And: *Please, Dad, please, how is your heart?*

After the most recent surgery, he sat in my dining room and we talked genetics. Did my grandparents have heart problems? Would I? What about my son, Caleb, then six years old and looking at a picture book, some kid-friendly anatomy thing. "Grandpa," he said, pointing. "Is that what a heart looks like?"

Dad bent over the page. "Well, real hearts are more purple than red," he explained. "There are veins and tubes and blood—" I watched the two of them, heads together. You can tell they're related: same looks, same laugh, same fearlessness. I like thinking of the beauty we pass down, as opposed to the danger. "Tell you what," Dad told him. "When I get back to Alaska, I'll send you a box of deer hearts and you and your mom can dissect them."

A month or so later, Grandpa's box arrived by FedEx. It contained three frozen hearts, a few rounds of caribou steak and thirty pounds of halibut that according to my father is the world's best eating fish. I stuffed the freezer, leaving a heart out to thaw, and went outside to ride scooters with my kid. "By the time we come inside," I told him, "the heart will have melted and we can cut it up!"

He looked at me from underneath his helmet and said, "Why do we want to do that?" which, frankly, is an excellent question, one rarely considered by this little boy specifically and, I wager, by children in general. Usually they jump right into that strange neon puddle. They run headfirst into that electrical fence. They grab that scorpion. They eat that thing. They stick their fingers in that other thing. And who the hell cares why?

I'll tell you who.

Your mother.

But with the arrival of the deer hearts, our roles had switched. The child now requires a logical explanation before he'll go anywhere near the creepy, drippy things, and I find myself both unable to stay away and unable to articulate why.

*

The next time I dissected a heart, I came prepared. I studied anatomical diagrams. I read websites on coronary artery disease that terrified me and articles on cardiology that were way over my head. I watched YouTube channels on dissection, videos of open heart surgery, and Cristina Yang performing her first solo valve replacement because oh my god I love that show.

Left atrium, chordae tendineae, ventricle.

Interventricular septum, tricuspid valve, papillary muscles.

I remembered the frogs I dissected in high school, skin flaps peeled back and pinned to the dissection tray; salmon yanked out of the Pacific, smearing blood from their hooked mouths and flopping like crazy till you whack them with a baseball bat; my friend Jeff writing about killing a fish as a child, how it influenced his understanding of queerness and masculinity; my son catching his first big fish on a river in Alaska, so excited, so proud, and his uncle Tim telling him he had to kill it, me standing there thinking of Jeff, thinking of the memories we carry; flash forward twenty years and my boy's grown up and still traumatized from killing this fish right here/right now; flash back to the present when he tells Uncle Tim no thank you and we go in search of skipping stones and talk about death: plants and fish and humans, scales and guts and bone.

I set the cutting board on the floor and called my dad.

"How's the writing?" he said.

"Ugh," I said.

"That good? What are you working on?"

My dad is one hundred percent supportive of my work. One notable and admittedly deserved exception: a published story where I spelled Dall sheep d-o-l-l and he went off, something about a Tom Wolfe novel where the main character goes quail hunting with buckshot. Buckshot! Can you believe it? Does he want to eat the bird or explode the bird? I mean if you can't even bother to research. I tell this story every semester to my writing students: Dall and doll, Wolfe and quail. We talk about movies set in Chicago that get our streets wrong. We talk about our responsibility to the places and people we write about. We talk about the pressure of telling stories about the people we love. "Are you scared to write about your dad?" they ask. And I answer, "Hell yeah I am!"

"I'm writing about you," I told him.

"Oh," he said.

I talked about the deer hearts, a metaphor for fear. Body, bones, blood. "You can read it first," I said. "In case you—"

"Oh, kid," he said. "Write whatever you want."

A tidal wave of gratitude.

"So long as it's the truth."

*

When Marilyn, my stepmother, called to tell me that Dad had been airlifted from the urgent care center in Glennallen to the hospital in Anchorage, I said, "Fuck," but with four or five *u*'s so more like: *fuuuuck.* I have a mouth like a sailor, but I try really hard not to curse around Mare because I know she doesn't like it and I love her very much. She gave me three brothers (we got drunk by the fire pit and decided we didn't like *step*), three sisters (we got drunk in the banya and decided we didn't like *in-law*), a brilliant niece (Hi, Olive!), a brilliant nephew (Hi, Nico!), and here we are, a family.

I remember the exact moment when I fell in love with Marilyn, not as my dad's wife, but as a separate individual that I'm lucky to have in my life. We were shopping for shoes somewhere on Michigan Avenue—Nordstrom or Bloomingdale's. The salesman unwrapped boxes and sat across from us. "Where are you from?" he asked Mare.

"Alaska," she said.

"Ha!" he said. "Sarah Palin!"

I am not Alaskan. I cannot and do not speak for Alaskans

and Lord knows I'm no fan of their former governor, but: Dear Lower Forty-Eight: That shit's getting old.

Marilyn smiled. This was not her first rodeo. "Are you from Illinois?" she asked the salesman.

"Chicago, born and raised," he said.

"Remind me: how many of your governors are in jail?"

Four—out of the past seven. Plus thirty aldermen and some fifteen hundred others—that's not a typo, fifteen hundred—convicted in the past forty years of bribery, extortion, tax fraud, embezzlement.* It was early 2009, immediately following the presidential election, and Marilyn and I had been talking about politics. We talk about politics a lot: abortion, education, gun control. To say that I am left and she is right is incorrect and oversimplifies us both, so I'll leave it at this: even though we see some things differently, it always feels like I've learned from her as opposed to fighting her. In our country's current mess of a political discourse, I walk away from our conversations thinking, *It's possible.* She's patient. She listens. She loves my dad and she loves my son and on the phone that day, her husband on a helicopter and scared out of her mind, she pretended I didn't swear. She knew I needed to say it.

Maybe she needed to hear it.

"I'll come," I said, reaching for my laptop. How fast we hit autopilot: call your boss, call your babysitter, book your flight.

"Talk to your dad first," she said, explaining that he'd call me before they took him into surgery. Like most Alaskans, she's pragmatic: flights are expensive, travel is complicated,

---

* Shane Tritsch, "Why Is Illinois So Corrupt?" *Chicago* magazine, December 9, 2010, http://www.chicagomag.com/Chicago-Magazine/December-2010/Why-Is-Illinois-So-Corrupt-Local-Government-Experts-Explain/

worry when the time comes. I am a midwesterner: we worry always about everything. I paced my living room, waiting for the phone to ring and imagining the worst. I knew Dad had been at moose camp, my uncle Chuck's setup on the mainland, but I'd never been there. In the absence of information, we substitute what we already know, so I pictured my dad on Barometer Mountain on Kodiak, its two-thousand-some elevation gain spread with wildflowers and ridiculously amazing views. He's wearing camo overalls, 7mm rifle at the ready, eye on something four-legged, almost has it, almost there, and then—a sort of tingle, like firefly wings on the inside of your skin, running up his arms, down across his chest, circling around his heart like a washcloth in a fist, squeezing tighter, tighter, body locked, and all you can see is sky. Look up: the ceiling is all clouds. So white. So close. The inside of your skin.

Chuck and my brother, Thomas, were there to help. The clinic in Glennallen sent him by medevac to the hospital in Anchorage.

Again.

For surgery.

Again.

*

My third dissection was a bust. I got out the knife. I got out the cutting board. I got out the charts I'd printed from the Internet and reached for my heart.

I don't know about your freezer, but ours is a god-awful mess. My dad brings us fifty-pound boxes of vacuum-sealed fish and steaks wrapped in butcher paper, and then stands in front of the open top freezer of our city-apartment-size fridge, perplexed. He forgets that we don't have a second chest freezer in the basement or the garage. He forgets that we don't have a basement or a garage. Then he gets this very determined look on his face like: *Dammit they're my kids. I have to help them.* And then he takes everything out of our freezer and puts it all back in with great strategy and expertise, like how secret agents in the movies build bombs inside suitcases. The result is a hundred-pound wall of meat intricately stacked inside 6.08 cubic feet, and everything else is jammed in the inside of the freezer door.

What happens next is life. You pull out one salmon filet and the stack comes crashing down. You make juice popsicles that spill before they freeze. The cap to the vodka doesn't get screwed on all the way. Ziploc bags don't get zipped. Opened bags of frozen peas upturn. Food goes bad. There are . . . smells, and somebody starts collecting animal hearts for reasons not yet understood.

That is the freezer I expected to see.

What I saw instead was very clean, bleached white, organized, and nearly empty: two ice cube trays. A pint of Ben and Jerry's (flavor: Americone Dream). A neat, small stack of fish. A neat, small stack of steak. No hearts.

I texted my husband. Who else would've been in the freezer?

Where is my heart?

A beat or two passed.

?

The deer heart. That Dad sent.

oh i cleaned the fridge

He texted in all lowercase, sans punctuation.

I see that. Where is my heart???

I used three question marks so he'd know I meant business.

He texted back with five.

?????

The conversation overlapped, thumbs flying.

THE DEER HEART.
i threw out a ton of stuff
IN A ZIPLOC.
it was going bad, freezer burn
I NEED IT.
YOU CAN'T JUST
THROW MY STUFF OUT.
wait are you actually pissed
YES I'M PISSED.
why
why
i don't understand why

I didn't understand, either.

*

My dad hunts big game, and is gone weeks at a time tracking moose, elk, caribou, whatever license he drew in the lottery run by the Alaska Department of Fish and Game. Sometimes it's a family thing: Dad, Chuck, my cousins, and their friends setting up camp, hauling gear in four-wheelers, cooking over open fires, navigating trails in twos or groups. But sometimes he's alone at twenty-five-hundred-feet elevation with just a backpack, a rifle, and, to keep his daughter happy, an iPhone. When he first moved to Alaska, not long after I left for college, he built a boat in his backyard: twenty-eight feet of welded aluminum with a bed, a GPS, and a Dickinson diesel heater. It took him three years. He had to learn how to weld. He's always been into fishing and now he takes people out on the ocean, the ones who come to Alaska for a quote *wilderness adventure* end quote. Before that, he was the elementary school principal for the children of the Coast Guard, and before that, he was the middle school principal in Chelsea, Michigan. He was *my* middle school principal in Chelsea, Michigan.

If you have not had the pleasure of your parent administrating school governance while you are going through puberty, here are a few fun anecdotes.

I was waiting outside my dad's office. A girl walked out and we locked eyes. I could tell she'd been crying. Later, at the football game, she held me down by the neck under the bleachers and told me she was going to shave my head. I remember the thick of her palm on my throat. I remember our team must've done something great because everyone above us stood up and cheered. I remember wondering, randomly, *when* she was going to shave my head.

I remember the fear—can't move, can't cry, breath locked.

Fast-forward to adulthood: This same girl sent me a friend

request on Facebook. I've been staring at her name in my inbox for a year.

My freshman year in high school, I'd walk the ten or so minutes back to Dad's office at the middle school to get a ride home. There were girls, a couple years older, who followed me every day in their car. They'd go like five miles an hour, all of them leaning out of the front and backseat driver's side windows, staring at me. I kept my eyes straight ahead. To this day, I'm not sure who they were, let alone what I did to make them mad. I wish I'd known. I wish I'd stood up for myself in some spectacular way, telling them off or keying their car or taking them on at the ALL-VALLEY KARATE TOURNAMENT and when my leg gets hurt in the semifinals my awesome guru coach uses an ancient pain-suppression technique in the locker room and the jerky asshole coach tells my opponent: *Sweep the leg, Johnny!* But I do a crane stance and we're all friends in the end.

That didn't happen, of course.

I kept walking. I pretended I wasn't scared.

Later that same year, in gym class, I hit one of those girls in the face with a softball. It wasn't on purpose. I was just shitty at sports. But still, I was sure she would kill me. After school, I asked my dad for help, and we drove back to the high school, its open, outdoor campus weirdly designed for California, not Michigan. I thought he was going to talk to the gym teacher, but instead we went to the baseball field and threw balls till it got dark. Can you picture him, still in his suit and tie, slamming a fist in his glove and trying, through sheer force of will, to summon forth some sort of athletic ability from his awkward, acne-ridden, bookwormy dork of a daughter? In my memory, we were out there for hours. We were out there for years. Hell, we're out there today. "Power doesn't work without aim," he told me. He'd said the same thing when he taught me to shoot, the two of us meandering brush lines, low-growth forests, and farm fields in search of a partridge, a pheasant, anything to aim for, to work toward, to hit, a final effect after all this cause.

*

"You're doing what with deer hearts?"

I'm drinking martinis with Randy, the chair of the creative writing department where I teach. This is the man who gave me: *Love in the Time of Cholera, Last Exit to Brooklyn, and Memoirs from the Women's Prison* by Nawal El Saadawi. He took me to Sister Spit to see Dorothy Allison. He took me to the Goodman Theatre to see *The Odyssey.* He took me to The Dragon's Den to hear Stanton Moore. He taught me how to listen. He taught me how to teach. He was the first person to show me that my writing had value and, in the same breath, challenge me to make it better.

"I'm dissecting them," I told him.

"I see," he said. "Why the hell are you doing that?"

I tried to explain: blah blah metaphor blah. Randy waited patiently as I talked myself in circles, finally arriving tipsy at the truth: *I'm afraid he will die. I'm afraid of the mountain. I'm afraid for his heart.*

"Have you told that to him?" he said, which in retrospect is a very good question but at the time seemed insane.

"You can't just say it!" I said.

"Sure you can. It takes"—he counted on his fingers—"five words." He studied me for a second and said, "For you, a thousand." He grinned. "Maybe ten thousand."

It was nice to see him relaxed. It was nice to talk about something besides the college. "You want to find a job that's not a job but a calling," he'd told me when I first started teaching there, and it certainly felt that way: radical pedagogy, diverse student body, what bell hooks called education as the practice of freedom. Then the new corporate leadership: faculty jumping ship, fired or forced out. It wasn't a job but a heartbreak.

We had another drink and he told me about cutting up cows on his family's farm in Minnesota. Ever since I started this thing with the deer hearts, everyone wants to talk about meat. About butchers. About dissection and hunting and organ donation and blocked arteries and invasive surgery—our battered, aging bodies, so beautiful and mortal. I love these stories, how one opens the door for another.

We paid the check and got our calendars, trying to schedule a meeting for I don't remember what. I suggested the following Friday morning, but he couldn't. Something about a doctor's appointment. Some tingling in his chest. Nothing to worry about. Everything is fine.

That Friday afternoon he had emergency quadruple bypass surgery.

I blame the college.

I blame the mountain.

Crazy or not, it's easier with somewhere to aim.

*

I texted my father, asking if he could send more hearts.

How many? he wrote back.

One would be great!

Or two!

Five?

My phone rang.

He explained that he was out of hearts and of course he'd love to shoot me a deer or two or five but he couldn't because deer weren't in season. Question: *"Does it have to be deer? Would—"* he paused to consider his words. *"—this thing you are doing work with caribou? How about elk? Sheep?"*

I didn't know.

I didn't know what this thing I was doing even was.

The next day I bought pig hearts from a butcher in Rogers Park. They were smaller than the deer hearts; at home on my cutting board they looked like valentines. I sliced through the veins, white and thin like dental floss, and remembered: the frogs I dissected in high school, where they'd come from and how they'd died; my son's uncle Lott and uncle Ryan taking him to an organic farm in Indiana, playing with pigs and chickens and cows that would later be slaughtered; waiting until the pheasant took off into the sky, the barrel of my rifle following it into the sky; waiting on the table in the paper robe, a doctor's fingers on my moles; waiting on a different table in a different robe, a doctor feeling for the cyst that swallowed my ovary; different table, different robe, a doctor taking blood and blood and blood; different table, without the robe, naked and open and terrified.

"Mommy?"

I looked up. My son, now eight, had just come in from swimming. Raw meat was piled on the counter. I wore blood-smeared rubber gloves. My laptop was open, playing a video of open heart surgery, the live organ pumping, sucking, and stuck with needles.

"What are you doing?" he asked suspiciously.

The pause was long.

Finally: "I'm writing an essay."

He eyed the meat. "That's an essay?"

I tried to explain, a task since Montaigne. "It's a kind of question," I finished.

"Okay," he said. "Did you find the answer?"

*

When I was little, my dad and I spent weekends together. We'd drive the green station wagon to a very specific field with a very specific tree. It had a heavy branch, like an arm, that stuck out from its side and ran parallel to the ground. Dad wore head-to-toe camo, waiting for birds to hit the sky, and I'd sit on that branch and bounce, watching our dog disappear in the underbrush and then leap above it, like he was riding waves.

There were always guns, but they were in my periphery; this was me hanging out with my dad. I was too young for a hunting license. I could shoot, but not to kill. I liked target practice. I liked shooting skeet. I liked the kick in my shoulder, the clay pigeon exploding midair and the tiny hole in the center of the target, proof that I got what I aimed for.

"Great shot, kid!" my dad would say. That's the part I liked most of all.

At some point, I stopped. Sundays were for handbell choir. I played the high bells and could pull off both a four-in-hand and a triple Shelley. And Saturdays were *Casey Kasem's American Top 40,* which I'd record on my boom box—yes, boom box—and play nonstop for the entire week. Look: I was almost a teenager. I was, like, busy. I had, like, friends. We went to Briarwood Mall and did *I don't know, nothin'*, which in the late eighties meant Hot Topic, Wet Seal, and Orange Julius, and now, I think, is H&M, Forever 21, and Starbucks.

That's what I told my dad, anyway.

The truth was more like this: I'd just taken a mandatory hunter's education course. We watched a slideshow on conservationism and Michigan-specific regulations; then we went outside for shooting, blood trailing, and tree-stand safety. I knew most of it already from weekends with my dad. But now it would be different. Now it would be about killing.

Dad paid the six bucks for my small-game license and we went to our field. I remember passing the tree where I'd bounce when I was little. I remember Duchess scaring up a pheasant. I remember watching it climb, reaching an almost fixed point in the sky and then—freeze. A perfect shot.

I so desperately wanted him to be proud of me.

*

"Do you remember when I dissected frogs in high school?"

We're in Alaska for the month of August. It's sixty degrees on the island, twenty hours of sunshine on the solstice. There's fishing and hiking and laughing. My son plays with his cousins in an enchanted forest that Marilyn made in the backyard, complete with teeny furniture for fairies. My husband drinks beer with my brothers, laying in the grass to measure distance between bocce balls. My dad paces the living room, phone at his shoulder, calling his neighbors to find me a heart.

"It's weird," I went on. "I don't remember the teacher, or even what year it was, and you'd think I'd have a problem cutting animals because I was a vegetarian in high school. Right? And you took me upstairs and opened the chest freezer and pointed at the frozen venison like: *This is the food that we eat, Megan. What food are you going to eat?* And I—"

"Hold on," he said to me. And "Maybe?" into the phone. Then he covered the receiver and asked if buffalo hearts were okay.

I nodded.

"Buffalo's fine," he said.

Long pause.

"My daughter's in town and she needs them."

Another pause.

"For art."

*

After I stopped hunting with my dad, he started giving me books. He played me Paul Simon and Lyle Lovett. He bought us season tickets to the Wharton Center in Lansing where the Broadway touring companies perform in Michigan. We'd get dressed up, drive the hour or so east, eat somewhere fancy, and off to the show. *Les Mis. The Phantom of the Opera. Cats*, the glowing eyes before the curtain rose. *Once on This Island*, the best singing I'd ever heard. *The Wiz*, the best dancing I'd ever seen.

I thought all of this art was for my benefit, a way to connect away from fields and rivers. But the more I get to know my dad as a person instead of just my parent, I can see that he needed it, too. In grad school, he'd studied the Concord writers—Thoreau, Emerson, Louisa May Alcott and her father, Bronson. When I was ten we packed the pop-up camper and went to New England. I remember the pencil scratches on the walls where the Alcott sisters marked their growth. I remember how still it was at Walden Pond. I remember my father's face as he stood at Thoreau's grave. Years later, I taught Kafka classes for a summer study abroad program in Prague and took my students to the New Jewish Cemetery where he was buried. I'd been in his head for years: journals, stories, letters. I have friends who've described this same feeling at Jim Morrison's grave, or Kurt Cobain's. Bruce Lee and Sinatra, Michael Jackson and Oscar Wilde and Princess Diana.

There's something in the wind. History crawls on your skin.

Someday, I'll take my son there. I'll ask him to describe my face.

*

A box arrives in the mail. It's from Alaska, an address I don't recognize. Inside are four frozen hearts and a note:

3 deer
1 buffalo
One of the oddest packages I've packed, enjoyable but odd.

I take a picture of the note and post it to Instagram. Then I get back to work.

Right marginal artery, diagonal artery, auricle.

Pericardium, pulmonary trunk, superior vena cava.

In case you didn't know—a deer heart has four chambers. So does a human heart. A frog heart only has three. I was surprised to find that fact in my head, twenty years after high school. Also in my head: cutting the frog open with scissors. How its tongue is connected to base of the mandible instead of back near the esophagus. Peeling away thin layers of membrane and finding its tiny, jelly-like heart between the shoulder blades.

Not in my head: What the classroom looked like. How the desks were positioned. My notebooks. My lab partner. My teacher.

Later, I wiped heart off my hands and got on Facebook.

> I know we haven't talked in years (hi, hello, how are you?) and I know this is totally random and probably insane, but I was wondering if you remember dissecting frogs in high school.

I sent this message to a handful of old friends, people I remembered fondly from show choir and debate. The responses went something like this:

> Great, how are you?
> Yes, this is insane
> YES I REMEMBER IT WAS SO GROSS
> I hate science
> Those poor frogs!
> I don't know . . . who was the teacher?
> I have blocked out high school entirely.

ugh, who cares

Wait.

Did Mr. Leith teach that class?

He taught chemistry, right? Did he do anatomy, too?

It was Mr. Leith, wasn't it?

Shit. It was Leith.

*

It was the end of my first semester of college in Boston. I was in the dorm, studying for final exams. Soon I'd be flying home to Michigan, the first holiday since my parents had split the summer before. Which one would I stay with? Where would I spend Christmas? I felt guilty for choices I hadn't made yet, selfish for wanting time with my friends, so desperately eighteen.

A girl from down the hall stuck her head into my room. "Aren't you from Chelsea?" she asked. "Chelsea, Michigan?"

My high school was on the news. There were the buildings, our ridiculous outdoor campus snow covered in mid-December, the baseball field where my dad and I played catch. There were cop cars in the parking lot. *School shooting,* said the newscaster. *Unknown fatalities.* And later: *Condition unknown.* Then: *Critical condition.* And: *No word at this time.* And: *Fatalities unknown at this time.* And: *One fatality at this time.* They didn't name names, of course.

Only this: *The shooter was a local schoolteacher.*

And this: *The victim was a local school administrator.*

A local school administrator.

This was before cell phones. Before voice mail and call

waiting. Before all of us on the Internet and Twitter with its to-the-second news. Back then, we had to wait. No one was picking up at either of my parents' houses and I couldn't get through on their work lines. I paced my dorm room, listening to the busy signal. In the absence of information, I filled in the blanks: a school board meeting maybe, one of countless I'd attended as the principal's kid. My dad at a conference table in his suit and tie. Mr. Piasecki, the superintendent, would be there, and Mr. Mead, the high school principal. I knew both these men. Were they okay? They had kids my age. Were they okay? Our families got together sometimes. Were they okay? Other people would be there, too: the elementary school principals, maybe some teachers and parents. Was the shooter already in the room? Did he burst in later? What happened? Where was my father, under the conference table, a desk, a supply closet—can't move, can't cry, breath locked.

We learn to fear our own imagination.

Five hours went by before I heard his voice. Five hours before I knew he was okay. Maybe that doesn't sound like long, but stare at a clock for a single minute and see how calm you feel. Shut your eyes and count to ten: one one thousand, two one thousand, three. Imagine what can happen to a body in that amount of time.

Imagine what can happen to a heart.

*

I don't want to remember Stephen Leith. I don't want him in my head and he doesn't deserve my heartbreak, but memory gives a rat's ass about permission. A smell can take us back. A look, a taste, a song. I hear Paul Simon and I'm eight years old in the car with my dad. Leaves turn colors and I'm twelve in a field with a rifle. Slice into a heart and there I am in high school, cutting up frogs. I remember the formaldehyde. I remember they were double injected to better see their veins. I remember the stuff: scalpel, scissors, forceps, pins. I remember the *ugh!* and the *gross!* and the *ewwww,* but the teacher is gone, wiped clean by time or anger. He was my teacher—that much I know—but everything else is constructed from media reports and legal documents, a story nobody wanted to be part of.

Some students liked him, apparently. He won a few teaching awards. He played in a band. He had a ponytail. He was thirty-nine years old, the same age I am now, and on December 16, 1993, he walked into the administrative offices at Chelsea High School with a 9mm Browning semiautomatic handgun. It wasn't a board meeting, like I'd imagined after hearing the news reports. It was a grievance meeting: superintendent, principal, and union rep. Leith shot all three of them, killing Joe Piasecki and wounding Ron Mead and Phil Jones before his wife, Alice Leith, came in and told him to give her the gun.

Mrs. Leith was my AP English teacher. She taught me Shakespeare. She taught me that "a lot" is two separate words. She prepared me for the ACTs and SATs, those vital steps to getting the hell out. Once, a boy in my class asked her about the purpose of literature—why did he have to study it, what impact did it have on his life—and Mrs. Leith gave an incredible monologue about language and poetry and what it means to be a human being.

This woman married to *that* man.

It's enough to break your heart.

When she asked for the gun, he set it down on a desk. Then

he went back to his classroom and graded papers until the sheriff arrived.

Later, it came out that he'd been reprimanded for "behaving inappropriately" to female students. Later, he'd say that medication affected his judgment, even though he was clear enough to reload. And later, years later, I'd read about his guns: eleven total, including an AK-47 assault rifle.

*

My dad wasn't at the high school when the shooting happened. He arrived just afterward, one of the helpers I'd learn about later from Fred Rogers. When I finally got through to him that day, when I heard his voice instead of the busy signal, my first reaction was relief. It didn't happen. And then, almost immediately, shame. It did. A man lost his life. A girl lost her dad.

"Here," I want to tell her. "Here is my heart." And I want those words to mean something. I want them to mean everything. *Our hearts are with the families,* we say and nothing changes. *Our thoughts and prayers,* say politicians who've taken thousands from the NRA. *Relief* that it happened to someone else, somewhere else, on a different block or neighborhood or community or country. *Shame* or *guilt* when we feel bad, and who has time for that? People are dying. That man should not have had eleven guns. That man should not have had *a* gun. His right to a gun is not greater than our right to walk through this world, alive and living.

"I want to be a helper!" says my kid.

Me too, baby. Me, too.

*

This past February was my dad's seventieth birthday, and he and Marilyn went to New York to see *Hamilton.* My aunt Sally and uncle Bob joined them and I showed up as a surprise, sliding up next to him in the impressionist wing of the Metropolitan Museum of Art. "What do you think?" I asked, referring to the painting in front of us. "Well—" he started, launching into his ideas, the colors, the form, not even looking at me at first. Long story short—hugging, laughing, drinking, and just after midnight, in the first few hours of seventy, my dad decided he wanted to join the crowd outside the *Today* show and try to get on TV.

"Don't people line up early?" Mare asked.

"Five in the morning," said Bob, who'd looked it up on his phone.

"We won't sleep!" Dad said.

"Who won't sleep?" Mare said.

"I need a poster," Dad said. "It has to say I'm visiting from Kodiak. People love Kodiak." He looked at me. "Do you have any poster board?"

That's how I found myself, two o'clock in the morning, running around New York City in the freezing rain looking for twenty-four-hour poster board. And markers. "Good markers," Dad insisted. "Not those thin pointy ones." We wound up in that pharmacy—what's it called—Rite Aid, which for future reference has a surprisingly well-stocked office supply section.

"Anything else?" I asked, my arms full of paper and scissors and nonpointy markers.

"Yeah, grab me some nitroglycerin."

Everything froze.

"In case I have a heart attack."

We stared at each other.

I could almost see him counting down: three, two—and he laughed.

He *laughed.*

So I laughed. We laughed our faces off in Rite Aid.

"You know I'm going to write about this, right?" I said.

"Why do you think I said it?"

Three hours later, he was on the *Today* show.

Three days later, he was back on the mountain.

*

Last night, we had a dinner party. Randy drove down from Oak Park. Jeff brought wine from the bistro where he works. Lott and Ryan played cards with my now eight-year-old son. Christopher prepped steaks for the grill; his hands were covered in sauce and he asked Randy to grab him something from the freezer.

"What's this?" Randy asked, holding up a Ziploc bag.

Everyone looked at me.

"It's my heart," I said. "My last heart."

It felt meaningful, somehow. The room was heavy.

And then, in one of those loud little-kid voices, my son said, "Let's cut it up!"

A plan was made: thaw the heart, eat dinner, and dissect. The table was set, wine poured. Randy talked about the farm in Minnesota. Ryan talked about the farm in Indiana. Lott talked about sheep in Florida. Christopher talked about agriculture class in Texas: "Mostly we watched movies about animal deformities." Jeff talked about his heart surgery: "I watched them put in my stents!" Randy talked about his: "I watched my angiogram!" And my son asked what the heck we were talking about. These are the men in my life. They love me even when—especially when—I act bonkers.

Naturally, we got drunk and forgot about the heart. But later, when everyone was gone or asleep, I got out the cutting board.

I remembered Alice Leith lecturing that kid in English class about the meaning of literature; Billy's mama running dog entrails under the faucet in *Where the Red Fern Grows*; Fleur Pillager working at Kozka's Meats; Raskolnikov holding the ax over his head; "I Am a Knife," by Roxane Gay; Sethe and her children and the handsaw; Materia cutting her daughter open with sewing scissors in *Fall on Your Knees*; Lidia holding her stomach in *Chronology of Water*; Janis Joplin screaming: "Take it!/ Take another little piece of my heart!"; humans as blood bags in *Fury Road;* Indiana Jones hanging from a rope bridge with Mola Ram; every episode ever of *True Blood*; the rhythm sequence from Jeunet's *Delicatessen*; Louise Erdrich: "How come we've got these bodies? They are frail supports for what we feel"; Charles Yu: "You want to tell a story? Grow a heart"; and Eileen Myles: "I can hold your heart for a second. And that's all anybody ever wants."

I've always engaged with the heart as a metaphor: a desire, a thing to survive, to heal from or shoot for.

Now I know there's nothing more real.

We walk through the world at its leisure. We're here at its mercy and with its blessing.

At some point, we have to ask ourselves how we want to live.

F

The summer between high school and college I worked nights at an Arby's in a truck stop near Dexter, Michigan: a cute, idyllic white-picket town, which to me was just a pee break on the way to Detroit. I mopped the floors of that pee break. I wiped the booths with Lysol after truckers ate their dinner. I stood behind the counter in my blue visor and took orders for Beef 'n Cheddar sandwiches, which I wasn't allowed to make because even though my birthday was in August, I was still a year too young to use the deli-style slicer that shaved roast beef. Question: Why is eighteen the acceptable age to operate potentially hazardous machinery? Is there a certain maturity I gained that summer between high school and college? Did I become more knowledgeable, like snap your fingers and suddenly you're this totally together person who can be trusted with industrial-size blades the same way that: *Snap*—ten years old and you can shoot a gun. *Snap*—sixteen and you can drive. *Snap*—twenty-one and you know how many vodkas are too many vodkas?

More than anything, I wanted something to happen.

I was invincible—on the edge of my life. Do you know that

feeling? Standing on a three-or ten-or twenty-story rooftop, leaning forward, the whole world's spread out before you, and you think: *God, what if?*

What if I jump?

What if I fly?

But I couldn't. I had three more months till I left for college and an eight-hour night shift.

Arby's sauce.

Jamocha Shakes.

Curly fries poured into wire mesh baskets and lowered into the deep fryer, spattering boiling oil like knife tips on my skin that hurt like holy hell but still felt good 'cause I was feeling something, anything besides my incredible seventeen-year-old need.

The summer between high school and college is supposed to be revelatory in some way, right? I pictured mine like a coming-of-age sort of rom/com dramedy involving the following: (1) cushy day job selling muffins at the farmer's market, (2) home to my cute, idyllic white-picket family, and (3) nights making out with the boy I loved. Note the wordplay: *boy I loved* as opposed to *boyfriend.* I knew he was seeing other girls, but during that summer between high school and college I equated love with danger and Lord knows, I needed some. Four years of A's, scholarship to a swanky school out East, goody-goody rep that was a total coping mechanism—all of which made my love for the boy I loved a total cliché.

I spread quilts on my bedroom floor to silence any footsteps, removed the window screen, and climbed two stories down a fairly precarious satellite antennae. The boy I loved was waiting at the lake behind my house, and we rowed out across the glass-like water, a million stars reflecting above. It was beautiful: mid-Michigan, mid-May, our bodies tangled in the bottom of the boat.

The sun had risen by the time I climbed back up the antennae, and as I struggled to replace the window screen I heard footsteps outside my door.

"Megan?"

It was my dad, and I dove fully dressed under the covers and held my heartbeat. He'd heard me. He must have heard me. Their room was just below mine. "Get downstairs," he said through the door, and I was not invincible then. I was the opposite of invincible. I was fucked.

We sat in the living room, a place reserved for the most serious conversations. I counted squares in the Oriental rug and wished my parents would say something, anything, please anything besides: "Did you have sex in the rowboat?" So when they told me they were getting a divorce, it's fair to say I'd asked for it. I remember they sat on opposite sides of the couch. I remember forty squares in the rug. Did I cry? Did we talk about how love isn't rowing around under the stars or twenty-six years of marriage? Did we talk about how no one is ever invincible?

I don't remember.

What I do remember is that neither of them moved out, and the tension in that house was so heavy I thought my bones might crack. I did everything I could to avoid it: making out with the boy I loved, staying over with friends, wandering zombie-like through the aisles of a twenty-four-hour Meijer's until finally deciding that I might as well earn some extra money. And so:

Arby-Q sandwiches.

Philly Beef 'n Swiss.

Curly fries dumped frozen into wire baskets, dipped into oil, wait till they cool; wait till the end of a shift; wait till I could leave for college and jump off the ledge—1:00 a.m., 2:00 a.m., and after a while, standing over that deep fryer with its boiling wire I started to imagine what would happen if I pressed the inside of my forearm against the basket, connecting my skin with the sizzling crosshatch.

That would be something, certainly.

Three months is not a long time, but that summer—a fucking glacier, a watched fucking pot. I'd look at the clock above the food line, shocked that only a few minutes had passed since the last time I'd looked. To pass time, I counted onion rings and wished for something, anything, please anything to happen, so when the boy I loved came in to tell me about the other girl and how maybe she might be pregnant, it's fair to say I'd asked for it. I remember we sat across from each other in the shiny vinyl booths. I remember the smell: Arby's in my hair, Arby's under my fingernails, Arby's up my nose and pores. Did I cry? Did I say awful things, trying to pick a fight 'cause if he was standing there yelling at least he'd still be standing there?

I don't remember.

What I do remember is chicken tenders.

Horsey sauce.

Curly fries dumped into boiling oil. It was August, two weeks left and I could jump: new life, new people, new heartbreak, and for the thousandth time I wondered what it would feel like to lay my forearm against the basket, to solder my wrist to the wire, to smell my own flesh burning through to the bone, to be an active participant in my life instead of responding to what was happening to me, to do something, to do something.

So I did.

My arm and the wire were in contact for only a second, but that's all it took. I screamed and stuck my left arm into the ice bin, still holding the sizzling basket of fries in my other hand. Coworkers must have found me there. They must've brought me to the Emergency Burn Station. They must've asked what happened, because I remember saying it had been an accident. I'd been careless, I told them, even though hurting myself had taken all the care in the world.

Later, at freshman orientation, I'd wear long sleeves to cover the bandage and, later still, a four-inch leather cuff to hide the scar, a purplish rectangle of chicken wire that would, over time, fade to a horizontal line bisected by two smaller lines. In other words—an F.

F is for fear.

F is for fly.

F is for French fry.

Later, I'd get a job as a food runner at a fancy southwestern restaurant. Cheese sticks, stuffed mushrooms, jalapeño poppers in wire mesh baskets, boiling oil like knife tips on my skin. I couldn't believe the want.

And later, after I'd left that college and another; left the country and another, I'd find myself in Chicago in a bar with a woman I didn't know how to love. When she reached to pay for my drink, her long sleeves pushed back, revealing faded scars that climbed from wrist to forearm. That night was the first time I talked about what I'd done, how I didn't know why I did it, and how now, twenty years later, I still can't explain the pull toward danger. There is no: *Snap*—twenty-one and now I understand it. *Snap*—thirty-one and now I have healed. *Snap*—thirty-nine and here I am, invincible.

There's only me, on the edge of my life.

The whole world is spread out before me.

God, what if?

Stand Here to Save Lives

In a session on casting in Jennifer Peepas's introductory film course, she brings in an enormous file of head shots, including actors of all races, genders, sizes, and ages, and she instructs her students to imagine the characters they might inhabit. "What kind of story or genre do you think of when you see this person?" she asks. "Is there a specific role or type that comes to mind?" The excitement is visible. These future filmmakers know movies; they've been shooting them in their imaginations for years. "He's the secret agent," they'll say, holding up a photo, more often than not, of a white guy. "The superhero, the good cop gone rogue," and various other examples of the main character. Then, other photos: "She's the mom!" And: "He's the terrorist!" And: "The funny best friend," "The love interest," "The drug dealer."

That's when the discussion starts.

"What makes that guy the main character?" Peepas asks. "Is she anything besides a mom? Why is he a drug dealer?"

The discomfort is immediate. Students squirm in their seats. You can see their brains working, realizing the racism and sexism behind their responses, thinking of excuses, getting defensive, and just before it erupts, Peepas flips the conversation from blame to responsibility. She gives them the data, how year after year, movie after movie, they've seen the same stories, the same characters, the same dominant narratives that privilege the same identity groups. "However," she tells them—and the future filmmakers lean forward—"you are the ones who can change it. You are writers, producers, directors, and media makers. It's on you to tell new stories, many stories, and to challenge the idea that there's only one way to represent a person or a people."

There's more to it, of course. Throughout the semester, Peepas works to build an inclusive classroom space. She brings in films and scripts from a wide variety of artists. And she considers how teaching the elements of her discipline—in this case: scene, structure, story, character—might also inspire a different kind of learning moment, the kind that saves lives.

"The first time I did this exercise, I thought it was just about casting," Peepas says. "But casting is never just about casting. It's all a teachable opportunity."

*

For years, I worked in a teaching and learning center at an arts and media college in Chicago. It's where I first met Peepas, along with countless other educators working to engage students in discussions about race, class, and gender-based oppression. They assign texts from multiple voices and perspectives. (You'd think this would be a given in 2015. Depressingly, you'd be wrong.) They prioritize community building, not just as a Day 1 icebreaker, but rather a necessary component in creating what educator Ken Bain refers to as a "natural, critical learning environment." They experiment with activities, assignments, and approaches that inform students of identity-based inequity in ways that are discipline specific, pushing back against the too commonly held idea that as one faculty member recently told me, "If students want to learn about race stuff, they can go take a cultural studies class."

A few examples of the many possibilities: A biology teacher using the Tuskegee syphilis study. A dance teacher using the Kennedy Center's recent performance of *Swan Lake*. An app development teacher bringing a mapping project that highlights places where women have made history. A journalism teacher inviting everyone, on the first day of class, to introduce themselves with their name and gender pronouns, opening a conversation with emerging media professionals about how to write ethically about identity as well as setting the expectation that the classroom is an inclusive space where everyone's humanity is respected.

Systematic oppression exists across all fields; the arts, the sciences, technology, the humanities.

Why on Earth would we not fight it that way it in the classroom?

*

Perhaps it's idealistic to think that what happens in a classroom can make a dent in identity-based violence and white supremacy. Perhaps some people think discussions of systematic oppression should be relegated to a single class (or a single chapter). Perhaps the deck is stacked too high against teachers: faculty of color are ridiculously underrepresented and often face hostile classroom environments, especially female faculty of color. Many are swamped trying to fulfill learning outcomes set by administrators who haven't been in the classroom for decades, if ever. Some are locked into teaching assigned syllabi with no say in the curriculum. Some have their programs gutted or outright cut. Non-tenure-track faculty, unsure of their jobs, are hesitant to rock the boat in fear of it reflecting negatively on their student evaluations. Adjunct faculty—fully half of the country's teaching force—are swamped trying to piece together a living wage, often teaching at multiple institutions and without time to attend unpaid trainings. All of our hands are tied as we wait for strategic planning, curriculum committees, and corporate administration, notoriously slow. The current cultural dialogue around higher education deems it—at best—broken.

And yet, every week, we walk into classrooms full of young people, ready to learn.

*

In a creative writing class nearly two decades ago, my fellow students and I showed up to an empty room, the tables and chairs shoved out into the hall. It was . . . weird. We were like, the hell is this? It was near the end of the semester and we'd been busting our asses on final rewrites. Exams were coming up in other classes and most of us had jobs, often more than one, to cover rent and tuition and unpaid internships and maybe food? Also: the pressures that come with future plans and familial expectations and what am I doing with my life? Also: student loans and a shitty economy and continuing job loss. Also: social lives, the ebb and flow of love and loss. Also: the quiet, individual mountains we were each trying to climb:

sickness and sick parents and single parenthood, violence and healing, addiction and recovery, and the daily instances of sexism or racism or homophobia so prevalent and relentless it's a wonder we haven't set the walls on fire.

All this to say: we were on edge.

Our teacher came in and told us that today we'd be taking a break. Today, we would just talk. "Today," he said, "we'll try to remember why we're here in the first place." We'd been with this man long enough to trust the process. His class was what safe space really means, allowing us to push past comfort zones into difficult work and difficult discussion. He asked us questions. It went like this: "If writing is your art, stand against that wall," he said, pointing to one side of the room. "If writing is your job, stand against the other wall." Then he'd indicate the empty space between and give us permission to stand there, as well—imagine a spectrum as opposed to a binary—whatever best illustrated how we felt in that particular moment. Then, one by one, he invited us to explain why we'd chosen to stand where we were standing.*

The discussions were incredible. We dug into definitions of art, of audience, what it meant to be a working writer, submission and publication and the editorial process, and the contributions we hoped our work would make to a greater cultural dialogue. We listened to each other, sometimes physically moving across the room based on what someone else said.

Our teacher asked, "What is missing from this conversation?"

---

* I would later learn that this activity was a variation on "the Spectrum" as developed by Guillermo Gomez Pena and Roberto Sifuentes and explained in their wonderful book *Exercises for Rebel Artists: Radical Performance Pedagogy.*

He asked, "What steps do you have to take to get from where you're standing now to where you want to be?"

He asked how we thought our work would be perceived. "If your writing is political, stand against that wall." He pointed. "If it doesn't have anything to do with politics, stand against the other wall."

I didn't have to think about it—not political—and when it was my turn to explain, I said, "I write love stories."

Across the room, an openly gay student was backed against the opposite wall. "I write love stories," he said.

Next to him stood a young woman of color. "Me, too," she said.

To this day, I struggle to explain what happened in that moment. All of the clichés apply: lightbulb, lightning, ton of bricks. I'd just turned twenty, from a very small, very sheltered town in southeast Michigan, and while educating me and other white, straight students was most certainly not these students' job, the simple gift of their perspective cracked the world open. It was the first time I'd considered how a person could be perceived differently based on their identity.

My discomfort was immediate. I felt my cheeks turn red, overwhelmed with shame for all I didn't know. But truly, who has time for such things? As Justin Campbell wrote at the *Los Angeles Review of Books:* "People are dying out here; we don't have time for bullshit." My teacher stepped in, flipping the conversation from guilt to responsibility. He gave us data. He assigned readings from multiple perspectives and invited us to develop our own.

He said, "You are the ones who can change it."

He said, "Where do you want to be standing?"

I walked across the room to the opposite wall.

*

In faculty development workshops, I shove tables and chairs into the hall. "If you're here to teach your discipline, stand against that wall," I say, pointing. "If you're here to save lives, stand over there." I point in the other direction and then indicate the empty space between.

Sometimes, teachers give me a look: *The hell is this?* Others have worked with me long enough to trust the process. But always, I'm grateful for the willingness to share.

In the decade that I've been asking this question—at institutions both within the academy and the community, graduate professors to kindergarten teachers, at educational conferences across the country—teachers have spread out across the spectrum, a bar graph of bodies. They explain their frustrations, their fears, the lack of resources and institutional support.

And yet.

Years ago I listened to Jen Peepas talk about an activity she wanted to try in a unit on casting. I remember she stood at the exact center of the room: teach the discipline, save lives.

We let ourselves dream.

What would that look like in my classroom?

What would it look like in the world?

twenty, or Good Lord, It's Me, Jane.

12

I picked the littlest puppy. She had a leaky eye. We named her Duchess and she slept in my bed and did tricks: gimme that paw, gimme the other one, hold a cookie on your nose until we say go and then flip it in the air. I'd throw sticks into the lake behind the house; she'd leap off the dock, paddle around, and bring it back, tail wagging, so proud she'd found it. One time a boy I went to school with—I remember his real name but will here call him Dickwad—went out in our rowboat, catching turtles in a bucket. Later, he stood at the end of the dock and threw them in the water for Duchess to fetch. She swam and swam and swam but the turtles, of course, were already gone. She came back exhausted and defeated, tail between her legs, like she'd let us down. And Dickwad laughed. He laughed. He *laughed*. I felt something in my hands then, climbing up my arms and into my chest. I wanted to punch him. I wanted to attack, to shove him to the ground and pummel him, my sneakers in his groin and stomach and face. It's the first time I remember wanting to touch a person in violence. I could imagine myself doing it, could feel his bone crack under my fists, his skin stuck in my fingernails.

13

Occasionally I meet other adults whose mom or dad had been their school principal. We share a look. Sometimes we high-five. Then we tell each other how the stoner kids our parent put in detention regularly locked us in our lockers, ruining marijuana for us to this very day and we all know that's the real crime here.

13

Let's talk about sex education. In public schools. In the eighties. In the United States of America.

We were separated into two different rooms. The girls went with the nurse, and the boys went with the gym teacher in the knee-high athletic socks and short shorts and keys on a very long string. Disclosure: it's possible that gaps in my memory are filled in with every television show ever made about high school. Was there really a gym teacher with knee-high athletic socks? There must have been. There always is! He's loud and tall with hairy arms and he takes the boys to their own room and shows them—what? What did you learn in that room, guys? Erections, right? STDs? Sperm meets egg? Did you learn about women's bodies? I'm especially curious about the boys who grow up to become government officials. What about consent? My research says no,* and if that's the

* Marisa Kabas, "We Asked Men How They Learned about Sexual Consent," *Fusion*, June 13, 2016, http://fusion.net/story/313091/how-men-learn-about-sexual-consent

case, I'm interested in how you first came across the concept. Last week, a college student came to my office hours. He had a question about the assigned reading, a personal essay by a woman about sex. He was wondering, I mean, he knew it wasn't my job and all, but could we talk about what consent, like, looked like? He'd never heard the word. He wanted to be sure he was doing it right. He wanted to be a good person, a good man. He wasn't sure who to ask.

I was floored.

I thought about the conversations I have and will have with my son. I thought about conversations in my classrooms, at colleges and universities regarding sexual assault policies, in the media about rape—Cosby, Ghomeshi, Brock Turner, and the powerful letter Turner's victim read aloud at his sentencing, which in my opinion should be required reading for all men and boys, for all women and girls. "To girls everywhere," she writes, "I am with you. On nights when you feel alone, I am with you. When people doubt you or dismiss you, I am with you. I fought every day for you. So never stop fighting, I believe you."* Jesus, we need to hear that, whether we're sixteen or sixty. We need real conversations about our bodies and our rights, in our culture and our literature and those terrifying girls-only rooms with the school nurse watching videos.

Like this:

A cartoon girl, drawn vaguely to resemble a bubble with

* Katie J. M. Baker, "Here Is the Powerful Letter the Stanford Victim Read Aloud to Her Attacker," *BuzzFeed*, June, 3, 2016, https.//www.buzzfeed.com/katiejmbaker/heres-the-powerful-letter-the-stanford-victim-read-to-her-ra?utm_term=.fw4oDwverv#.wnnwVxgMyg

lips and yellow pigtails, comes running out of a door marked BATHROOM and says, "Oh no, I'm dying!"

A cartoon rabbit hops over and says, "Why do you say that, Judy?"

"Because I just went to the bathroom and I'm bleeding!"

Big blue cartoon teardrops shoot out of her eyes.

The rabbit chuckles. "You're not dying, Judy! You have your PERIOD!"

The word PERIOD appears on top of the screen in big bold letters.

"My period? What's that?"

Cut to a diagram of the female reproductive system. Bright red arrows run through squiggly lines connecting blobs marked OVARIES to a triangle marked VAGINA. "You're becoming a woman, Judy!" says the rabbit, hopping across the diagram, not unlike those Energizer Bunny commercials.

Fade out.

Or this:

Fade in to a wide shot of a racetrack. Lined up at the starting gate are hundreds of cartoon sperm looking eerily similar to the minnows Michigan children catch in bogs and light on fire. One smiles big—a toothpaste commercial star appearing on his white teeth—and waves at the camera. We know we're rooting for him because he's the only one with a defined face. Then a gun fires, and—they're off! The track is a long pink tunnel and the minnows swim top speed, little bubbles coming out of their mouths. Some get too tired and stop, or they bang into unseen obstacles and disappear. Our sperm, though, pushes admirably on. We're excited. We cheer, as if we've put money on this sperm and the payoff is 5:1. Soon, our sperm is the only one left and we see what he's been swimming toward: a white cartoon egg, exactly like the one you eat hard-boiled for breakfast except with a face: big red lips and lashes over her eyes.

"Hi!" she says to the sperm.

"Hi!" he says.

Cut to a real baby, happy and cooing in a fluffy pink dress.

Roll credits.

Or this:

LOOK AT THIS PICTURE OF GROSS HORRIBLE DISEASED GENITALIA. THIS IS WHAT WILL HAPPEN IF YOU HAVE SEX BEFORE MARRIAGE. BOYS WILL THINK YOU'RE A WAD OF CHEWED-UP GUM. NONE OF THEM WILL MARRY YOU. YOU'LL BE ALONE FOREVER WITH BOILS AND PUS ALL OVER YOUR PARTS. I CAN'T SAY THE ACTUAL WORDS FOR YOUR PARTS BECAUSE THOSE WORDS ARE OBSCENE AND MAYBE ILLEGAL AND ALSO I DON'T KNOW THEM. I KNOW THEY ARE DOWN THERE. SOMEWHERE. WHEN YOU ARE MARRIED YOU WILL ONLY HAVE SEX WITH THE LIGHTS OFF SO IT'S FINE.

End scene.

I was mystified by these videos. My mother had always talked to me about my body in ways I'd later understand as feminist: truthful, anatomically accurate, body positive, and body autonomous. I knew that I would discharge blood and mucosal tissue from the lining of my uterus and though that sounded horror-movie terrifying, it was actually normal and no big deal. I had a starter kit for girls under the sink, individually wrapped squares of Advil in my backpack, and a smart, well-educated adult woman at home with whom I was comfortable talking about puberty and sex. If you were one of the many, many parents who avoided these conversations with your own daughter and relied on the public school system to fill in the blanks, I was one of the many, many girls who found her crying in the bathroom after choir practice, promised her she wasn't dying, and taught her how to use a tampon.

That said, just because I knew what was up didn't mean I wasn't awkward. The day I got my period—oh my god, this is mortifying—my mom made my dad take us out to dinner in Ann Arbor. To celebrate. She gave a toast: blah blah miracle of womanhood blah and I died a quiet death. To be clear: my dad was totally cool. His whole career involved squirrely, preadolescent kids running around with hormones oozing from every pore. The embarrassment I felt was mine and mine alone. Had it been possible, I'd have set off a time bomb to avoid further discussion.

15

I played Winnifred in the Chelsea High School production of *Once Upon a Mattress.* I couldn't sing for shit but I did shout in a sort of lyrical way, plus I had what is referred to as "personality." If you're unfamiliar with the script, Winnifred's first scene happens after she supposedly swam across a moat, meaning that immediately before I walked onstage, the stage manager dumped a bucket of water on my head. I'll leave it at this: I was not wearing the right kind of bra to appear cold and dripping in front of my entire small-town community.

Also: during tech rehearsal I was goofing around and accidentally broke Prince Dauntless's foot, so the super-cute senior playing him had to perform the whole show from a chair.

15

January 17, 1991. I was in the high school auditorium for some rehearsal or another, the musical or show choir, forensics or debate. I loved all of that stuff. I loved being a part of something: those first messy steps of collaboration, learning how to create, to make, a song or a dance, a scene or an argument. There were ten or so of us sitting stage left, and someone came in from the side entrance and said that the United States had just bombed Iraq.

The bubble of my town, my high school, my family, my privilege. I was scared of Scantron tests; scared of Mr. Terpstra's class, reciting vocabulary words in unison as he pounded on a desk: *i-ron-y a state-ment or e-vent in which the opp-o-site is said or the un-ex-pect-ed hap-pens;* scared of dissecting frogs in biology class, the smell and the death and those tiny jawbones, so ridiculously fragile compared to the deer carcasses hung upside down and bleeding out in my garage; scared of the older girls who followed me in their car; scared of walking onstage and forgetting my lines; scared of college applications; scared, like thousands of small-town kids, that I'd never "make it out of here"; scared I'd disappoint my parents; scared of the choreography to "Cold Hearted"; scared boys wouldn't like me; scared girls wouldn't like me; scared no one would like me; but that night in the high school auditorium, on break from whatever rehearsal had brought us together and listening to the news about Iraq—it all seemed small.

It was the beginning of an ongoing dialogue I have with myself about my own privilege.

16

As soon as we got our drivers' licenses, we'd go to Ann Arbor, a ten-minute drive east with movie theaters and shopping malls and the University of Michigan (read: college boys). In the summer, they play outdoor movies on top of a parking structure not far from the Power Center, this towering auditorium made of mirrored glass. We'd bring blankets and snacks and stay out until *midnight*. ("The movie doesn't even start until ten, Mom! It has to be, like, dark!") One night, a friend brought a friend who was dreamy as hell and he asked if I wanted to sneak into the Ray Charles concert.

If you are in need of a pickup line, that one totally works.

Of course, we couldn't get in. The Power Center was packed, and two dumb kids were no match for security. Instead, we lay in the grass just outside those huge glass walls and from there, watched the stars and listened: "Shake Your Tail Feather" and "Singin' This Song for You" and "Hallelujah I Love Her So"—muffled, of course, but still. Later, we wandered to central campus, sat on the edge of the *Ye Gods and Little Fishes* fountain and stuck our bare feet in the water. We must have talked; I don't remember what about. I don't remember what I was wearing, or how he looked at me, if we touched, if the heavens burst open or time stood still or "We are young/Heartache to heartache we stand," but as far back as I reach down the line of my life, this was the first time I was ever in love.

What's more terrifying than that?

16

I was one of those special kinds of geeks who cut class to hang out at the library. I read Tolkien and *Franny and Zooey* and *Sassy* cover to cover and Sylvia Plath way before I knew what she was talking about and, in 1991, the news. That was the year the world cracked open. Years before we all searched Google a hundred times a day, I took my questions to the librarian: Desert Storm, Duvalier, Mandela, Kevorkian, Rodney King, Exxon, Croatia and Slovenia, Mike Tyson, Anita Hill, Khmer Rouge, HIV. I pounded my poor parents with questions and kept exhaustive notes in my journal. We should all be in awe of teenagers, of youth, youth artists in particular. Holy hell, the emotion! The love and the anger and the energy, all so huge, enough force to power a city. I think back to myself then, and I look at the young writers I work with now, and am blown away by their courage. It scares people, I think. We try to contain it. We teach them to hold back. To be "appropriate." To be "respectable." I wonder: *What might happen if we got out of their way*? What might happen if we actually listened?

16

High school was complicated. I imagine that's the case for most people. Where do you put the frustration? The hormones? The fear? Sometimes, at night, I'd spread blankets on the hardwood floor to muffle my footsteps and sneak outside to the lake. Keeping your balance in a boat is tricky in the dark, but I'd lie still as stone and float across the water, feeling like I was inside the Hubble Space Telescope. You've seen this scene in dozens of movies; the hero looks up, millions of stars blanketing her from above, and has some sort of epic epiphany about how tiny we are in the grand scheme of things. How connected we are in the grand scheme of things. How trusting we are in the grand scheme of things. In high school, I fiercely believed in grand schemes, and those stars seemed like proof, an endless bibliography for my desperate teenage questions.

Let's be honest: I wasn't just sneaking out to row around and *have feelings*. If you're unfamiliar with southeast Michigan, there are many small lakes, and on any given night you'll be able to find a couple of teenagers rolling awkwardly in the bottom of a boat. I remember lying on my back. I remember the stars. I remember the whispering: *Shhhh somebody'll hear us. Shhhhh be quiet!*

I didn't want to be quiet.

If you head west from downtown Chelsea, toward the big lake where the actor Jeff Daniels lives, there's a curve of train tracks that cuts across the road and runs parallel with it for a half mile or so. And then, right before the road and track split again, pulling away from each other like a slingshot, there's a little shed. To this day, I don't know what it's for—maintenance supplies, maybe? It's where I'd go to get away, at first riding my bicycle and later, after I learned to drive, my awful orange Toyota, the one I'd gas up with meticulously collected pop cans cashed in at Polly's for ten cents each. I'd sit on the little steps leading up to that shed. It would take the longest time for my eyes to adjust to the darkness, allowing me to make out the outline of my own hand in front of my face. In the rare moments cars would pass, I'd go temporarily blind from the brights turned on high against the deer regularly crisscrossing the road, but then they'd be gone and again—darkness. And then—stars. I'd wait there, sometimes a few minutes, sometimes an hour, but eventually I'd hear the train, first a dull roar from miles off and then louder. I could feel its tremor, climbing up through the tracks and into my shoes. I could see its headlight illuminating everything like a stage spotlight, coming closer, faster, louder, the engine about to eat me alive and finally—finally—I screamed. That's what I'd wanted to do all along. Scream my head off. Throw my voice at it, all of it, the frustration and the hormones, the confusion and the doubt, everything so raw and wonderful and terrifying.

16

I'm neckdown underwater—can't move, can't cry, breath locked.

I look up and standing at the edge of the quarry are boys. Maybe they're men. I can't tell. The day had been so beautiful, the water so warm, blue green and glass-still in this dugout mineral pool where we'd come to make out and skinny-dip in the sunshine. He'd left for some reason—food, drink?—I don't remember the specifics but it would only be for a moment and that was fine, more sun for me. I floated on my back, hair splayed mermaid-style around my head, listening to my breath underwater, in and out, in and out, and suddenly they were there, first just one and then he called for the rest. In my memory there are six or seven, but maybe there were only three? Four? How many would it take? They stood at the edge of the jagged rock, looking down at me trapped in a fishbowl below them. Instinctively, I locked myself into a ball and moved toward shallow water, low enough so I could touch bottom but still high enough to shield my body, to cover myself, to hide.

Our relationships with our bodies are complicated. I have not always treated mine kindly; other times I'm blown away by its beauty and strength and capacity for joy. This moment in the water was the first time I remember fearing it.

"Stand up," they yelled. "We want to see you!"

How much time passed?

"Come on! We're not gonna do anything!"

Five minutes? Ten?

"Are you fucking deaf?"

"Bitch, stand up!"

"We're not going to fucking rape you!"

We're not going to fucking rape you.

In the end, the boy I was with came back, and the others left. This was one of a thousand moments in the girl character's life in which something could have happened, but didn't.

16

If every film about high school is to be believed—*Carrie, Heathers*, and *Pretty in Pink*, for starters—nothing is scarier than prom.

Mine was incredible. I went with my friends Casey and Jeremy, the twins who didn't get the Doublemint commercial. They showed up on my doorstep wearing matching tuxedos and brought their friend Dave, who'd made his tuxedo out of newspaper. I had a black dress and fishnets and three count 'em three hot dates, so basically I'm in debt to these guys until hell freezes over. A week later we snuck off to the dunes at Lake Michigan, sleeping bags in the sand. At some point I woke up to the moon like a spotlight above us and, on my right, Casey, rocking his head in his sleep and, on my left, Jeremy, rocking his feet.

A year later I came home from Massachusetts for spring break and met up with Casey at his dorm at the University of Michigan. We went to the Law Library, this ridiculously beautiful Gothic cathedral, and laid in the grass. The stars were incredible; it had been so long since I'd seen them. Casey told me he wasn't happy in Ann Arbor. I told him I wasn't happy in Boston. I don't remember what else we said, but by June, we had both left those cities, those colleges, those lives.

Twenty years later my family and I were with Jeremy and his three-year-old daughter at Kalahari, an indoor waterpark in the Wisconsin Dells. We'd booked a two-room hotel suite, thinking the adults could talk in one room and the kids would crash in the other, but they were too worked up, chlorine and sugar and spaz. I lay in one bed, singing Big Star to my six-year-old son: "Won't you let me walk you home from school/ Won't you let me meet you at the pool." Jer lay in the other, singing *Grease* to his daughter: "Summer days drifting away/ To, uh oh, those summer nights." I adore this memory; our voices tangled in the dark, our lifetime of friendship, a model of men who are good and kind.

It's bigger than the quarry. It's bigger than the fear.

17

I was on the debate team. Senior year—regionals, I think? The topic was the environment. I argued the negative, which meant defending the status quo. The environment was fine. No global warming, no extinct species, no damage from oil spills, and as a very high and mighty vegetarian I didn't know if I could pull it off. Then I walked into that competition and saw all the boys. Hundreds of them, no joke. They wore suits. They organized their index cards alphabetically. They looked at me. At first I wondered if I had food on my face. I wondered if my skirt was too short. Then, as I came out of the bathroom, I overheard one of them say, "Looks like someone's here to make quota."

I'd like to tell you I kicked their asses, but honestly I don't remember.

It pisses me off that they're still in my head.

17

I didn't know how to react to my parents' divorce. I had yet to begin the lifelong work of figuring out how to fit between them. *If I do this, will I remind him of her? If I say that, does she see him?*

The essayist Chloe Caldwell writes: "I didn't think I had the right to be sad."

Here's what I do know: they loved me fiercely. I also know what a privilege that is: parents who love you, and who demonstrate that love in word and action, through childhood and adulthood.

17

I was scared that someone I knew would see me in my Arby's visor with its stupid fucking cowboy hat logo. Now, I'd like to reach back across time and give myself a little talking-to about what really matters: pride in hard work, saving for college, and standing alongside kind people who made me laugh.

18

New city, new coast, new college: new ridiculously tiny dorm room in Boston. I don't remember much about my classes. Mostly they were lecture-style, two-hundred-plus students sitting in auditoriums, the professor waaaay down there. We copied notes off of a slide projector. We took blue-book exams that were graded by TA's. I dated one of those TA's. He was nice but I didn't love him, so I saw other guys, too. That's what we called it—seeing people—which mostly meant checking out their CD collection and making out in the dark. I remember lots of groping. I remember wondering why nobody turned into princes. Was something wrong with my kisses? Was I the wrong kind of girl? We don't question the pop songs. We question ourselves, and I was mad at Whitney Houston and Lisa Loeb and all their heartfelt longing, right up until I walked down the hall in my dorm and heard a guttural, wailing sort of scream-singing with just a bass beneath it. I stood there, listening, finally knocking on the door and asking what the hell is that. The girl who lived there showed me the album cover: PJ Harvey topless in black and white and flipping her wet hair. The next day, I went to the Virgin Megastore on Newbury Street and bought the CD, eventually scratching it and going back for another copy. And another. Another.

It was the first music that felt like mine.

My favorite part, then and still: "I've called you by your first name/Good Lord it's me—Jane!"

18

I had one friend at college. Her name was Meg. Meg and Megan. She had glitter eye shadow, a fake ID, and Rollerblades. I had *Blood, Bread, and Poetry* by Adrienne Rich and a PowerBook 140 with a trackball and internal floppy drive. We cut class and went dancing and she got me a job at a fancy southwestern restaurant around the corner from FAO Schwarz. I was too young to serve alcohol so instead I was a food runner: meals came up in the kitchen and I carried them to the table. Back and forth, back and forth, all night, every night—and during the day, still as stone in auditorium seating, some professor or another reading from his PowerPoint.

To pass time, I counted jalapeño poppers.

One, one hundred, one thousand.

How many jalapeño poppers equal a semester of undergraduate tuition?

There have been many times I've been afraid of math. Mr. Clarke's AP algebra class. The miles between Boston and my mother. Buying a condo, losing a condo, setting up my kid's college savings plan, my own student loans, my husband's student loans, my students' student loans, and our country's student loans, but nothing is as awful as doing something you hate to pay for a waste of your time.

18

Journalism was not for me. My teachers said I used too many words and also there was this new thing called the Internet.

Show of hands—who's arrived at college and by the end of the first semester you're like: *What the hell am I doing here?*

I called my mother and told her I was switching majors. Something more realistic. Something with a future. "Oh, honey, that's wonderful!" she said. She's totally supportive, even when I do things that are bat-shit crazy. "Did you do your plus and minus thinking?"

I most certainly had.

"And?"

"I'm going to major in *philosophy.*"

The day before, in ethics class, we'd started Aristotle. Three hundred eighteen-year-olds and Aristotle. We sat in this giant lecture hall, the professor talking in front while TA's ran up and down the aisles with microphones. I watched the one I was dating. He wore very tight jeans. He had a very nice ass. I was not the only one watching him. I wondered if I was the only freshman he'd invited back to his apartment.

The discussion that day concerned the true nature of the self. "Aristotle wrote that one's actions define one's true self," said the professor, down at his podium.

I wrote ACTIONS in my notebook.

"As in, a knife's true self would be defined by cutting."

I wondered if a knife could have a true self.

"So consider," the prof went on, "what actions define your true self."

Hands shot up around me, and my TA handed the mic to a guy in John Lennon glasses, which in 1993 were most definitely a thing. "Professor," he said, his voice oh so academic. "Shouldn't we first define truth?"

A personal essayist spends a fair amount of time with this question, much of it in our own heads: *What really happened? Why can't I remember that part? Who was the guy in the blue shirt? That was a great shirt. I bet I could find that shirt online.* And so on, ad nauseam. I also find myself occasionally in the position of answering this question publicly, as though there's a succinct definition. As though philosophers and poets haven't been tangled in its nuance for years.

Sometimes, I talk about emotional truth, how Gregor Samsa didn't want to go to work so he turned himself into a bug. I'd wager there are people reading this right here/right now who really, really don't want to go into the office today. How big is that want? Can you turn yourself into a bug? Into a unicorn? Into a cloud and float away? I think those ideas are beautiful. I think they're profound. I think they're true.

18

There was one class I liked—Italian. It was small, only twelve students. We knew each others' names. We talked: dreams and plans and questions. We rode language like a wave.

18

I stayed in Boston that summer to work at the restaurant, and on our days off, Meg and I and our friend Joy went to Martha's Vineyard. We'd catch the Peter Pan at South Station to Vineyard Haven, then a ferry to the island. Joy let me borrow her purple-tinted sunglasses and I remember looking out at the ocean, the sky, the sand, thinking: *This is a postcard. This is a deodorant commercial. Everyone should wear purple sunglasses. Everything is perfect.* I didn't know such a world existed outside of the movies: sparkling ocean, multimillion-dollar beachfront houses, spit-waxed convertibles, pink-and-green cardigan sweaters tied loosely over tanned shoulders and shoes that cost twice my rent, bodies in bikinis and board shorts throwing Frisbees, and picnic baskets specially made to keep white wine cool in the sun. We existed at the edges, crashing in the woods with island squatters living in tents. I wore nylon grandmother slips from the Value Village Thrift and pretended I was in *Cat on a Hot Tin Roof.* We hitchhiked from beach to beach and showered in the ocean. We were always salty. We were always sunburnt. We were always high. The first time I did mushrooms, I lay in the sand with Joy's lap as a pillow, watching clouds make shapes.

"Why are you crying?" she asked. Tears dripped into my ears. Somewhere nearby, Meg was dancing. Meg was always dancing. Someone had a drum. Someone always had a drum.

"I'm scared," I whispered. "What happens when this ends?"

19

New country, new language, new school: an international study abroad program in Florence, Italy. Churches and castles and villas, angels on the ceiling and demons on the roof. Here were the real-life versions of what I only knew from books: the Duomo, Ponte Vecchio, and Boboli Gardens; *the Birth of Venus, Last Judgment*, and Michelangelo's *David;* Dante, Boccaccio, and Italo Calvino. Every day between school and home I'd walk down Via Roma, looking at the store windows with their beautiful dresses and beautiful shoes. I never went in. Ten years, I promised. In ten years, I'll be back. By then I'll have money. That's when you have money, right? Twenty-nine? Thirty? I'll walk into those stores like Julia Roberts in *Pretty Woman.*

It was 1994. The exchange rate was 1,569 lire to 1 US dollar.

Italy just barely lost to Brazil in the World Cup finals and I was shocked to discover how deeply I could care about sports. I studied European literature, fighting with language and dancing with language and listening to language. I studied Italian politics and was terrified of how little I knew about the world. Silvio Berlusconi had recently been elected prime minister; I thought how silly it was that a media tycoon with zero political experience was suddenly running a whole country.

19

I'd been there two weeks when my American boyfriend showed up. We rented a flat at the outskirts of the city and suddenly I was Living with a Man. No rushing to make it home by curfew, no sneaking out the window when our parents got home early, no roommate walking in on you in the dorm—just us, in our Italian flat with the roaches and the grappa and our grown-up double bed. I'm an awful sleeper, then and still. I lay awake, watching the red digital numbers on my alarm clock thinking: *If I fall asleep right now I can get two hours. If I fall asleep right now I can get one hour and forty-seven minutes. If I fall asleep right now . . .* These days, I get up—write, read, binge watch Netflix, you know the drill—but back then I watched him sleep.

He'd be deep in some dream, eyeballs twitching behind closed lids, body heat notched twenty degrees. This guy slept hot; in my memory I see steam. He'd sweat, throw the covers and sheets to the side, splay out on his back in a big X across the bed. I'd trace his muscles with a fingertip: shoulder bone; biceps, triceps; his chest a number three lying horizontal rising up, then down, up, down. I'd press my palm into his skin, feeling his heart throb. I imagined reaching into his rib cage and grabbing it, a fistful of blood and pulse and goop like that scene from *Temple of Doom.* Mola Ram is chanting, summoning Indy's heart out of chest suspended across that fallen bridge. I actually tried it once, lying over him as he slept, my hand tensed like a bird's claw, nails digging into his flesh, whispering words I'd made up to sound like Latin. I thought: *If I want it hard enough.* I thought: *He crossed the ocean for me.* I thought: *If he wakes up and sees me like this he'll think I'm crazy and fuck—he'd be right.* It was crazy not to trust him. Crazy to be scared. Crazy to feel so alone next to someone you love, a million miles in the inches between you.

19

Three months late, alone in a bathroom stall in a hostel in Rome. I peed on the stick. I waited the three minutes.

Twenty years to the day and I still feel those minutes in my bones.

19

On a table at the gynecologist's office, naked from the waist down. The doctor told me to take off my pants and when I asked for the paper robe, he repeated himself slowly: "Pant-a-lon-i," and pointed at my legs. "Uhm . . . toga?" I asked, pointing at the white paper that covered the exam table. I was far from fluent. Maybe I'd used the wrong word. Maybe I'd asked the wrong question. Maybe this was yet another thing I didn't understand: the language, the culture, my body, my heart. The pregnancy test had been positive. The week since then was autopilot, one foot in front of the other: catching a bus back to Florence from a school field trip to Rome; telling my boyfriend, who climbed fully dressed into the shower, which seemed like a logical response; asking one of my teachers to help me find a doctor ("Ho dolori menstruali," I lied); walking through the door of what seemed like someone's personal apartment with its antique furniture, love seats and low light; and finally—the table.

"Sono incinta," I told the doctor. I'd looked up the word "pregnant" in my English/Italian dictionary along with the appropriate verb conjugation for "do not want." "Non voglio essere incinta."

He leaned me back on the table and got between my legs. I wished he was a woman. I wished I could ask questions. I wished my mom was there. I wished I had a paper robe. I wished it was out of me.

He said something in Italian.

More Italian.

More Italian, ending with: "Non sei incinta."

I pushed up on my elbows. "No?" I said.

He nodded and took off his gloves.

"But—" Wait. What? It was too fast, too confusing. I searched my Italian vocabulary and gave up immediately. "I took a test," I said in English, miming a stick with my thumb and pointer finger. I should have been relieved, but instead I was terrified—can't move, can't cry, breath locked. He hadn't been down there for very long. What if he missed something? What if I wasn't explaining it right? What if I was—I *was*—wasn't I? "It said pregnant," I told him.

His accent was heavy. "The test is wrong." Then he got up to wash his hands and said a long string of Italian ending with pantaloni.

I hadn't moved, still on my elbows, naked from the waist down, my knees wide. "Are you sure?" I said. "Sicuro?"

"Sono sicuro," he said.

More Italian.

More Italian.

More Italian, ending with a question mark and he snapped his fingers in front of my face. That's when I cracked, crying and exposed on that fucking table. "Non capisco!" I said, I don't understand what you're asking, I don't understand what's happening, I don't understand my body, I don't know who to talk to, I don't know what to do, I'm so, so alone, and I'm so, so afraid.

He picked up my pants from the love seat and laid them over my thighs. "I ask, do you want medicine?" he said, gently now. "Medicine for birth control?"

I nodded.

I closed my legs.

I put on my pants and went back to class.

19

In Italy I could buy alcohol. I bought a lot.

19

In Italy I cleaned houses. I cleaned a lot.

19

In Italy I wrote in a journal. I wrote a lot. I didn't call myself a writer yet. I thought you had to accomplish some unspoken, insurmountable list of requirements before you could use the word: publication, recognition, MFA. I remember hours in an Internet café, looking at college websites. Do you remember that pressure? That panic? Where do I go, how do I afford it, is my application good enough, am I good enough, as if the entirety of our lives could be decided just like—snap your fingers—*that.* I found a place in downtown Chicago that housed creative writing in its School of Fine and Performing Arts. It seemed so radical: to teach writing not as an offshoot of an English degree but as an art unto itself, like painting and music and dance. They offered me a fellowship for transfer students and voilà—deal sealed. Chicago! Nelson Algren! Chaka Khan! Lake Michigan! Which I loved from the Michigan side and could now have with a city attached! A city with museums and theaters and rock clubs and poetry and also I could take the Amtrak to see my mom and do my laundry and why it seemed easier to take my laundry on a four-hour train to Ann Arbor instead of a block away to the Laundromat is beyond me but whatever! I was nineteen! Also! A boy I knew lived in Chicago and maybe we'd end up together! Maybe it was fate!

It wasn't.

The city was.

19

Before I left Europe, I spent a couple of months backpacking by myself. Mostly I stayed in hostels but occasionally met up with people in bars or cafés who I'd travel with for a while, crashing at their flat or in their tent. I ended up in Cannes 'cause I wanted to see the film festival, not realizing that I needed, you know, tickets, or something to wear besides hiking boots and a giant backpack loaded with quite literally everything I owned: clothes, sleeping bag, knife—you could travel with knives back then—a few books, and my journal. Instead, I put on my bathing suit and went to the beach: white sand, turquoise water, the whole nine yards. There I met two French boys, a cute one who spoke a little English and a really cute one who spoke none at all. "He buy beer!" said the cute one, pointing at the really cute one, and the really cute one pointed at me and said, "Run, Forrest!"

After a mostly indecipherable conversation, I understood that:

A) the really cute one thought I looked like Jenny from the movie *Forrest Gump* or
B) he thought I was Jenny from the movie *Forrest Gump* (?) and
C) he wanted to buy me a beer.

And what *you* need to understand is:

A) I'm saying really cute but
B) what I mean is total fox and
C) I hadn't had sex in three months since
D) the American Boyfriend left and
E) I was heartbroken because
F) I loved him
G) even though it was hard
H) even though it was sad
I) even though I knew it wouldn't work.
J) It is possible to hold all those feelings at once.
K) Also

L) he was the only person I'd ever had sex with and
M) I didn't think I could have sex with someone I didn't know or
N) someone I didn't love but
O) I was wrong.
P) Yay, France!
Q) Drinking in the sun!
R) By the ocean!
S) In the sand!
T) Of course he could buy me a beer!
U) Or two!
V) Or five!
W) And by the time we were sloshed and back at his hotel, my skin had scorched red, my bathing suit firmly tattooed on my back. It was this black strappy number and when Total Fox got me out of it he said a lot of French very quickly and put cold washcloths all over me.
X) He was very sweet.
Y) It's a lovely memory.
Z) I'm so glad it's mine.

20

There was a flash summer storm when we drove into Chicago, Laura and I in her tiny beater and Heather in the U-Haul packed high with stuff from our parents' homes in Michigan. The three of us were fast friends, introduced to each other the month before and now bound together by: *What the hell is happening*? The rain was a faucet and the wipers were shit. As we crossed the Skyway, *Jesus Christ Superstar* played on the

cassette deck. Picture it: crawling through the rain for a h
mile over a steel truss bridge, wind whipping, lightning r
off in the distance but right fucking there, thirty-nine lash
and Jesus screaming in the background. It could not have be
more ominous. If we'd been the tiniest bit superstitious,
might have turned back.

But then? The Chicago skyline, lit up through the wet a
the dark and the bullshit of everything we were leaving behi

Como Questo

Here is a memory: not in my head, but in my bones.

I'm scrubbing the floor, deep red terra-cotta tiles that have to be cleaned with a special red terra-cotta stain sold in expensive tins the size of my palm. I was given the tin and the tiny brush. I was instructed to push aside the furniture, to pull back the rugs, to get down on my hands and knees. "Pulisce como questo," he said, demonstrating how I was supposed to scrub—precise, rough, inch by inch, room by room—and then he sat at his desk, supposedly returning to his very important work.

Except he didn't work.

He watched me.

Despite the miles I'd put between myself and Michigan, I was still the same girl, scared and ready and green as hell. I'd needed extra money so a teacher from my study abroad program helped me find side jobs: filing here, babysitting there, and a part-time housekeeping gig an hour outside the city. I took the bus through Tuscan hills, nose to glass. There were olive groves. Cypress trees. Whole fields of sunflowers. Drop-

off was a dirt road, then a hike up to a hidden hamlet of villas, all cobblestone and secret stairways. You could see the Apuan Alps in the distance. It was so beautiful. It was so perfect. There's no way it could be real.

It took four hours to clean the house top to bottom. Two stories. Three bedrooms. Covered loggia, dark wood furniture, windows thrown open to da Vinci–type views. There was always espresso, the expensive kind in teeny cups. There were always peaches, velvet and dripping. There were books, stacks and stacks, all hardbound and well loved. I'd dust the covers and flip pages: Italian, French, German, English, and other languages I couldn't then recognize. They belonged to the men who lived there. Three of them. They did research and wrote. I'd been told they were students, but they were all older than me by what seemed like a decade so I figured it might be doctoral work. I didn't know if one of them owned the place and let the others live there, or if they rented together. I could have asked—I had enough Italian for simple conversation and if they could read English, they could speak English, right?—but small talk felt off-limits, like a line had been drawn in the sand.

We said *ciao.*

Grazie.

La prossima settimana?

Si pulisce como questo.

I dusted books and wood and heavy frames around oil paintings. I washed dishes, wiped windows with vinegar, and pulled weeds between cracks in the cobblestone. I scrubbed the toilet and the tub and my reflection in the bathroom mirror. I swapped old bedsheets for new bedsheets and loaded laundry in the washing machine and hung it to dry in the perfect sun and for the most part? It was fine. I worked hard, was paid well, and had time in my head to conjugate verbs.

Except.

At first, I thought he wanted to make sure I was cleaning the way he'd told me to clean: his house, his cash, his rules. But as time went on, I knew that wasn't it.

I can still feel his eyes: my back, my thighs, my ass in the air as I scrubbed como questo. From that position—tabletop, it's called in yoga—my breasts dangled underneath my T-shirt, ratty and red-stained. My ankles were visible between sneakers and red-stained jeans. When I sat back on my heels to stretch my back, he watched the arch and curl. When I lifted my arm to roll out the kink in my shoulder, my shirt rode up my stomach, exposing a stripe of bare skin, and he watched, panning slowly up my body till he met my eyes.

He saw me see. He knew I knew.

Listen: this was twenty years ago. The logistics are fuzzy. Did he follow me from room to room? Was his desk placed at some strategic location with open sight lines of the whole floor? Where were the other two guys? Did they know this was happening, or did they just stuff their share of lire into my envelope and disappear into their books? Sometimes, I picture all three of them, sitting together on the couch and staring at me like I was the television: my back, my thighs, my ass in the air.

So much of memoir is crafting a self, stepping back inside our long-ago lives and seeing through their eyes. I was surprised to find how harshly I've judged this girl, down on her knees and scrubbing this man's floor. *Why did you let yourself be treated that way?* I've asked her, internalized misogyny feeding my own shame. *Why did you keep going back? Why didn't you tell him off?*

No way would I tolerate other women being treated like this.

Why do I allow it in my own goddamn head?

The Build Up To and Takeaway From

Spin

I spent my first month in Italy in a tiny cobblestone city called Siena, so beautiful I thought it was pretend. The streets were narrow and lined with churches and cathedrals and museums: every ceiling was painted with stories, every wall molded into gargoyles and cherubs. At its center was a football field–size square called Piazza del Campo where everyone met up, drank in the culture, and in my case at nineteen, sat alone with a bottle of cheap wine—*Ohmygod I could buy wine!*—and cried over the boy I left back in America, the stupid boy whom I very well may have forgotten a few weeks later after meeting a Carlo or a Paolo, but it had only been two days and I was heartbroken, drunk as only an underage American in Europe can be. Ten feet from me, professional opera singers rehearsed *The Barber of Seville* in the open air, soaring alto and contralto. It was so beautiful. The sky was all stars, a singular still shot or picture postcard, and I looked across the piazza and saw him.

You're drunk, I told myself. But even so, it was him, the American boy I loved so first-love desperate, showing up as if he'd heard me from the other side of the ocean. I watched him cross the square, rubbing my eyes in case there was some smudge sickness I'd picked up in customs. It was him, it was, heavy backpack and dirty clothes and so decidedly foreign, so not this picture. He looked up and—yes, the cliché—our eyes met across the piazza. Can you hear violins? I was on my feet, running, pushing past the Italians walking hand in hand under the stars. I didn't slow down as I neared him. I sped up faster even, and we reached each other in a frantic mess of limbs on limbs on lips and hugging, tight, bodies locked together.

That's when he said it, whispering in my ear as he spun me around and now, after everything that happened between us—a year of living together in Italy, the sex, the scare, other lovers and different lives—this is the moment I choose to remember: spinning in circles in that pretend city, my feet off the ground and trailing behind me.

"I crossed the ocean for you," he said.

That's love, right?

Right?

Girlfriend

Dave was almost thirty, I'd just turned twenty, and therein lay the problem: how would I get in to the places he wanted to go? Later, he'd buy me a fake ID from a girl named Tiffany who was five foot two with black hair. From then on, I slouched a lot and said "dye job" as the bouncer looked from me to the card—me, the card. I always got in. Dave said it was because I looked older, but I knew it was because of him. He was a musician and, so I thought, knew everyone sitting at every door at every indie club in Chicago circa '95—Empty Bottle, Double Door, Beat Kitchen, Bottom Lounge—and when we got to the Metro that night, he took my hand in his and held them both out to the bouncer.

"Hey, Mark," he said.

And Mark said, "Hey, Dave."

And Dave nodded in my direction and said, "Could you get my girlfriend a wristband?" *Wink wink*. I got a plastic bracelet proving I was twenty-one, and then a rum and Coke, and when Dave and I were alone in the crowd I said, "Girlfriend?" I yelled it, actually. We were in front of the speakers.

Dave grinned. A grinner, that one. "What do you think?" he said.

Here's what I thought: *Run*.

I liked Dave: the places he took me, the music he turned me on to, his bed and his body so free of mess and history, but the last thing I wanted was to be somebody's girlfriend. Girlfriends were for lettermen's jackets in high school. Girlfriends were left in Europe, waking up alone in an empty flat. Girlfriends were the phone call my roommate got in the middle of the night from a recent ex, all coked out, saying if he ever saw her talking to another man, he would find her when she least expected it, he knew where she lived and where she worked and all the dark alleys in between, and as she listened, my roommate slid down the wall with the receiver at her ear.

Vibrator

His name was Cal Buckton. I loved that name. It made me think of cowboys, which Cal decidedly was not. He was a psychologist dealing with violent schizophrenics and was always taking somebody down.

"How was work?" I'd ask every night on the phone.

"Fine," he'd say. "Had to take somebody down."

He lived in Boston, and I'd just started grad school in Chicago. We'd met through a mutual friend when he came for a visit and from there, it happened fast: BANG POW DO YOU SMOKE WEED BANG POW SEX ON A FUTON BANG POW LONG DISTANCE RELATIONSHIP.

I liked our arrangement. We were together without being together. No navigating time commitments, no social compromise, no When to Leave Your Toothbrush in His Bathroom. Phone sex was great; I bought a fancy vibrator. I got a ton of writing done—stories, essays, chapters, thesis. Think of how much time you spend trying to find your next love. Think of how much time you spend trying to find your next lay. What could you do with that time? What could you make?

"How was work?" I asked one night, several months in.

"Fine," he said. "I asked if I could transfer to Chicago."

It was over pretty quick after that, but I still have the vibrator.

It's a LELO.

I *highly* recommend it.

Love

When Jeff told me he was gay, I cried.

I had so much to learn.

There are so many different kinds of love.

Penis

Middle of winter, middle of the night, middle of the park. It was beautiful, lit up like a stage show from the street lamps, the snow seamless and clean like just-laid carpet. I stood behind Brad, his back pressed into my front, my arms around his waist with his penis in my hands, writing my name in the snow.

First, the *m*.

Then, *e*.

Then, some dripping.

"What!" I said. "There are three letters left!" Brad didn't say anything, just drooped in my palm. "Come on," I said to his back. "You had like six beers!"

We were friends, both of us seeing other people but even if we weren't, there'd be no chance. We didn't want each other like that, no invisible gravitational pull. It made honesty easy. We could talk. Earlier, at dinner, I'd gone into a whole big thing about Christmas at my grandparents' house in Michigan. I was the youngest, one of three girl cousins, and in my little-kid memory, the boy cousins numbered in the hundreds. In truth there were eight. Eight boys and one bathroom, so they'd go to the porch. They'd stand in a line at its edge. They'd say readysetgo, undo their belts, lift the fronts of their puffy winter coats, and pee in the snow. There was a lot of stream crossing accompanied by dire warnings not to cross streams, and pictures and messages and names in wide wet script. This was something they did together, a sort of family bonding that excluded me by anatomical default. I'd watch from the sliding door, nose pressed to the glass. Typically, those boys took me everywhere: fishing, hunting, sledding, target practice. I wasn't used to being left out.

"Come on," Brad had said, putting money down on the table, and we crossed Milwaukee Avenue to the park and its perfect snow. "Don't say I never did anything for you," he said, unzipping.

It was the first time I'd touched a penis in a context that wasn't sexual. The *m*, the *e*, and, soon after, the *g* and the *a* before he ran out of juice.

Zippers

I was visiting my dad on an island in the Gulf of Alaska. He had a job there at an elementary school on the Coast Guard base. One of his students' dads worked on a huge military vessel called a cutter and offered to give us a tour. It had living accommodations for the entire crew, out on the ocean bringing relief supplies to Indonesia and rescuing commercial fishing boats stuck in storms. The day they'd returned to port was the same day we boarded, so, in a nutshell: I was on a big-ass boat with a hundred-some guys, none of whom had seen a woman in months.

Back home in Chicago, I'd recently discovered one-night stands. I'd walk into Innertown Pub or Tuman's and scan my options: too old, too drunk, not enough tattoos and then, inevitably: *Hello, Danger.* He'd be skinny, already sloppy on PBR, bent over the pool table with his cue lined up on the eight ball and just before he'd let 'er rip he'd be compelled by some higher force—you might say the Lord but science calls it pheromones—to look up at me, and I'd look back at him, our eyes tying us together in the electric space between our bodies.

Take that look and multiply it times the Coast Guard.

The lower deck was a maze of tight, narrow hallways with low ceilings and low lights. Sometimes, as crew members passed in the opposite direction, you'd have to scooch up against the wall to let them pass, and sometimes, pressed up against that wall, you couldn't help bumping shoulders, or thighs, feeling wrinkles from waterlogged fingers, eyes like lasers, the rubbery vinyl of their zip-up waterproof jumpsuits. I MEAN COME ON! JUMPSUITS! You can't get easier access: reach up, pull down and whoop, there it is.

I was twenty-two years old, near stupid with want and fantasy. I'd been such a good girl. *Please, thanks, you're welcome.* So appropriate. So passive. Lie on your back, gasp, accept, and wash the sheets afterward but man, that cutter. Those bodies. Each one I passed was more delicious than the last and my imagination went bat shit, him and him and him and heat heavy in the icy air, vinyl wet and slick on muscle, every corner a new hiding place to unzip, unbutton, and bend over: engine room, storage room, pilothouse, stern ramp. I wanted all of them. Any of them. I wanted, for once and goddamn all, to be honest about my own desire.

Hands

I was with Josh, we were in a bar, fuck if I remember which, and he'd had a lot to drink 'cause work was like *fuck*, and his roommates *fuckers*, and apparently I'd fucked up, too? Whatever it was, it must've been awful, like, I must've gone too the fuck far because three seconds later, count 'em, one two three he flipped over his barstool, threw his glass on the floor, and reached for me, hands outstretched. I remember the veins in his forearms, purple rivers on a map, his biceps like baseballs. And people were on their feet, kicking aside their chairs, backing away from us, and I was backing away from him coming at me and he was yelling and I was still confused because what? Because why? But I didn't have time to put it together, the long line of cause and effect that started with a drink and ended with his hands, knuckles locked and tense and close and closer and just before they connected with my face, the bouncer caught him from behind, his meatball arms locking around Josh's neck and—

I was lucky.

A few days later, he would gently pet my forehead and say, "I never would've hurt you." At twenty-four, I believed him.

At twenty-five, I imagined inventive ways that he might die.

At twenty-eight, I learned how our past relationships can affect our current relationships.

At thirty-three, I began raising a boy to be Not That Man.

At thirty-nine, I'm ready to forgive myself.

Cock

He was sitting in one of those spinning office chairs. I was on my knees between his legs, nose to cock. I was supposed to be giving him a blow job, which can be amazing and sexy and hot as hell when things like consent and trust are involved but in this particular scenario, I was pretty sure that he was filming me.

We were in his studio, a garage sort of setup where he made experimental videos. That's what he wanted to do, anyway. What he did do was record weddings and children's dance performances and then sell the tapes. This was the early 2000s, everything everywhere was going digital. He didn't have money to upgrade his equipment, so he made quote art end quote. I was not there for the art. I was there for the sex, but man, I had a bad feeling about all of it: the way he'd positioned his chair just so. The care in how he arranged his body, like when you're getting your picture taken you suck in your stomach. How he kept looking behind me, over my head.

Fort

Todd's apartment was on the top floor of a fifth-floor walk-up. The stairs gave me plenty of time to think. "Can we talk?" I asked, and we sat on the step in front of his door. A lot of words came out of my mouth then. I don't remember exactly: something about trust. About how, in the past, I'd been careless with my body and my feelings and my safety. About how I like seeing you, Todd, but I need to know that if I walk through that door—I pointed at his door—I can still see you tomorrow.

He smiled and said, "I thought we could build a fort."

When was the last time you built a fort? The last time you had that kind of fun? The sheets and chairs and pillows, the impossible, puffy architecture? We turned his living room into a low-ceiling tepee, lying underneath with blankets and a flashlight. We laughed a lot. It was joyful. Easy. Safe.

The next day, still giddy and high off the possibility, I turned on my phone and listened to a voice mail he'd left about how we shouldn't see each other.

I'm still pissed.

He is the only one I was ever honest with at the beginning.

Hip

I was late. The band had already started by the time I got ID checked, hand stamped, and in through the side door. The bar ran the length of the wall to my left, stage at the far end, and everything in between was bodies: dancing, pulsing, banging into each other. I scanned the crowd, wondering how do I find—and Selly appeared at my elbow. She was tiny, her body delicate but double the attitude, tough-girl chic, tattoos and army bag and dyed red hair. We *hey*'d each other, hugged, and she made the *you want a drink* gesture. I'd had too much wine at dinner beforehand, but I am very good, then and still, at convincing myself that more is okay, more is healthy, I'd been working so hard lately, the meetings, the deadlines, that grant, and I needed this night, I'd nap tomorrow after finishing everything that needed finishing. I nodded to Selly and looked at the band. They were loud. Head full of noise and that was good. No room for bullshit if you're already full.

Selly came back with something on the rocks. As she handed it to me, her shirt rode up with the lift of her arm, exposing a smooth line of skin just above her belt loops. Her tummy was flat, hipbones curving out and up like sculpture. I couldn't stop staring. I gulped at my drink, ice on the heat between us. I wanted to throw the glass against the wall. I wanted to touch Selly's hip. I wanted to squat down on the floor, wrap both arms around her waist, lip to that smooth line. I wanted to stay there for the next hundred years.

All my life, I've looked at other women's bodies in envy. I don't like admitting that. I should be stronger, smarter, more confident, but fuck, here we are. Look at her ass. It's higher than mine. The skin under her arms doesn't flop. Her thighs are tight. But this thing with Selly—it was different. It was the first time I'd ever looked at another woman's body not in comparison, but in want. To touch it, lick it, blend it with my own.

"Hey." Selly leaned in to get close to my ear. "What's going on?" she said, but she knew. We both knew. We'd been living in this moment for months, had talked about it a thousand times, and it came down to this: I wanted sex, she wanted love, and neither could give the other what she wanted.

"I have to go," I said, which is what I always said: time to move, too much to do, that grant, those meetings, and sleep sometime, right? No time to think, or process, or handle, the drama, the desire, the buildup, building, higher, and then I was out the Double Door's front door, hand in the air, in the cab, off to the boyfriend's house.

It was so much easier with him. He didn't need me to love him.

Love to Watch You Go

He had the best ass I'd ever seen, very high and round. I'd ask him to get things—water, grapes, condoms—just to watch him climb naked out of bed and walk across the room. I'd think: *I hate to see you leave but I love to watch you go.* Then I'd laugh at myself for being so ridiculous. It's good to laugh at ourselves, I think.

One night, after sex, we were lying on the couch. Spooning, it's called. Me in the front and him in the back. "I'm in love with you," he said to my hair.

Then: "Did you hear me?"

Then: "Megan?"

Then: "Hello?"

And I said: "Can't we just watch a movie?"

Neither of us moved—no breathing, even. Then he sat up, untangled his body from mine, and walked across the room to his clothes. I watched him disappear into his jeans, shirt, and shoes, then out the door, then gone for good. I could have said something. I could have explained myself, apologized, talked about fear and trust, but in the moment? Being awful was easier than being honest.

Lightning

As a birthday gift, or maybe a gag, some friends snuck into Christopher's apartment while he was at work and covered everything with aluminum foil. It was a studio, not much space to cover but still: foil-wrapped clothes in foil-wrapped drawers. Foil-wrapped books on foil-wrapped shelves. Foil-wrapped individual aspirin inside the foil-wrapped aspirin bottle inside the foil-wrapped medicine cabinet over the foil-wrapped sink. It was my job to keep him away, and when we walked into the apartment, the doormat crunched. When we turned on the lights, we nearly went blind. When we fell back on his bed, it crackled beneath our bodies, papered us in silver and shredded as we flipped: me on top, him on top, me, him. Later, he peeled foil off the windows so we could watch lightning split the sky, the best show in springtime Chicago. We pulled the screens off, too, and stuck our feet in the rain. We fell asleep like that, two pairs of bare legs sticking out of a ninth floor window over Pratt Boulevard, half on the bed and half in the sky. It was perfect—magic, even.

It was my last first time.

When I look back at my sexual history—those singular still shots and picture postcards—so little of it involves the actual physical act. Rather the before and after—the buildup to and takeaway from. It's me figuring out what I want and what I'm worth, a long line of cause and effect that started with spinning and ended in electricity.

By morning, the room was a microwave, sun baking the foiled walls and cooking us from the outside in. But hey! Who gives a shit? We were in love. We drank cold coffee and unwrapped his apartment, book by book, dish by dish. It took days. We'd stop to sleep, to fuck, only putting on clothes to go to the 7-Eleven for ice and cereal and Gatorade. Every night, it stormed. So humid. So much sweat. So much sex. Even now, after everything that's happened—moving to Prague, eloping on the beach, near foreclosure, living off an art website and writing books and our perfect little boy who, blink and you'll miss it, just lost his first tooth—I can still remember how those first nights felt. I don't mean the scene-building stuff of here is his shoulder, here is my knee, the walls were painted red, it was nighttime. No, I mean the gut punch and thrum of muscle memory: electric shock on skin, wet and sticky and sliding, teeth in my neck, thighs taut, tighter, squeezing, holding, wait—and waves, unfolding, unlocked, and open and free.

This Essay Is Done

It was raining the day of Megan's funeral. I stood under my umbrella, looking at the tombstone with her name on it. MEGAN STIELSTRA *in big letters. It was very sad that I was the only person who showed up. She didn't have any family or friends that cared about her.*

*

This is the beginning of a short story written by a college student and turned in to his creative writing professor—me. He had turned in other stories as well, about how I had died. Sometimes, a character with this student's same name and physical appearance saved me from a violent perpetrator. Sometimes he was the violent perpetrator.

"It's fiction," he said. "Isn't that what this is? A fiction class?"

One day I walked into the main office and found him trying to talk a work aide into giving him my home address.

This happened fifteen years ago and this student is an anomaly among the brilliant, thoughtful writers I've been privileged to work with but man, I can't cut him out of my head, especially as colleges across the country debate the issue of Campus Carry. As of this writing, nine states enforce laws allowing license holders to carry concealed weapons into college classrooms, and it looks like more are on the way; a January 2016 report from the Education Commission of the States in partnership with NASPA concludes that while "numerous states prohibit guns on campus, the architecture and momentum of new policy represents a shift in the opposite direction."

FACT: Nine out of ten campus police oppose concealed weapons on campus.

FACT: Ninety-four percent of college faculty oppose concealed weapons on campus.

FACT: Ninety-five percent of college presidents oppose concealed weapons on campus.

FACT: Eighty percent of college students say they would not feel safe if guns were in their classrooms.

QUESTION: Would you?

*

At the University of Texas at Austin, students are fighting their recently enacted Campus Carry legislation with dildos (#cocksnotglocks). The short version is this: dildos are considered "obscene" and prohibited from campus so students are tying them to their backpacks by the dozens and showing up en masse at what student organizer Jessica Jin calls—wait for it—strap-ins.

I did a little digging.

It's always been easy to buy or sell a gun in Texas, but up until 2008? You couldn't buy or sell a dildo. And while there are no limits to the number of guns one may own, up until 2003 it was a felony to own more than six dildos. I found this information on my new favorite website—*dumblaws.com*—and confirmed it through: Texas Penal Code, Chapter 43, Public Indecency, Subchapter B, Obscenity. And since we're already in crazy town, here are other things that were illegal at some point or other: In my home state of Michigan, you couldn't swear in front of women or children. In my adopted city of Chicago, you can't have fuzzy dice or air fresheners hanging from your rearview mirror. In California, women couldn't drive while wearing a housecoat. In Alabama, you couldn't carry an ice-cream cone in your back pocket. In Cottage Grove, Minnesota, residents of even-numbered addresses could not water their plants on odd-numbered days excluding the thirty-first day where it applies.

But unregulated firearms?

Hey, have at it!

*

Over ten thousand people signed up via Facebook for the first University of Texas strap-in held at the same time that Campus Carry went into effect on August 1, 2016, which, in a stunningly offensive fuck you to basic human decency, is fifty years to the day that engineering student Charles Whitman climbed a three-hundred-foot tower in the center of the UT campus and opened fire from above, killing fourteen and injuring countless others. He used a .30 caliber carbine, a 6mm bolt-action Remington, a .35 caliber pump rifle, a 9mm Luger, a Smith & Wesson M19, a sawed-off semiautomatic shotgun, and seven hundred rounds of ammunition—all of which were purchased legally.*

It goes on.

On and on and on.

Whatever you believe about the Second Amendment, it contains the words "well" and "regulated," and in the last thirty years of mass shootings, licensed, legal, supposedly regulated guns are the norm, including the one used in 1993 by Stephen Leith in my hometown.

*

Remember: My father is an Alaskan, and a gun owner. He's also a Vietnam vet. He and my uncles and my cousins are lifelong hunters. I grew up in this life. Spare me the NRA horseshit that an AR-15 is used to hunt.

---

* Also legal:

The guns used to kill forty-nine queer people of color in Orlando.

The guns used to kill fourteen people at a holiday party in San Bernardino.

The guns used to kill nine people at Umpqua Community College in Oregon.

The gun used to kill nine black worshippers at AME Church in Charleston.

The gun used to kill four students in a high school cafeteria in Marysville.

The gun used to kill three people at Fort Hood.

The gun used to kill twelve people at the Washington Navy Yard.

The guns used to kill twenty-six people, mostly children, at Sandy Hook Elementary School.

Even the guns being used in record violence in my beautiful, complicated city of Chicago were purchased legally in states with less regulation: Indiana, mostly, but also Mississippi, Wisconsin, Missouri, Kentucky, Tennessee, Georgia, Ohio, Texas, Florida, Michigan, Iowa, Alabama, and others.

*

Recently, an emeritus professor at the University of Texas withdrew from his position teaching economics to classes of nearly five hundred. "The risk that a disgruntled student might bring a gun into the classroom and start shooting at me has been substantially enhanced," he wrote in his resignation letter. "I cannot believe I'm the only faculty member who is disturbed by this."

He's not.

If Campus Carry becomes law in Illinois, I'm out, too.

I can't teach wondering who has a gun.

QUESTION: Could you learn?

*

I want this essay to strap a dildo to its backpack. I want this essay to give those UT students a high five.

They're trying. I want this essay to try.

I want this essay to listen to its uncles and cousins about what they think common-sense gun legislation looks like. I want this essay to listen to why students and faculty might want to bring a gun into a classroom, to recognize those fears and come up with solutions that don't include weapons. I want this essay to sit on the floor of the House of Representatives behind Congressman John Lewis, protesting congressional inaction on gun-control legislation. I want this essay to put a spotlight on Mothers Against Senseless Killings and Assata's Daughters and the #LetUsBreathe Collective at Freedom Square and so many other community-based organizations led by black women working nonstop without resources or recognition to end violence in Chicago. I want this essay to go online right now and look up the names of its state representatives and see if they've taken money from the NRA and, if so, I want this essay to vote them the hell out of office.

This essay is done feeling helpless.

Nick Anderson, "If you want to carry a gun on campus, these states say yes," *Washington Post*, January 27, 2016, https://www.washingtonpost.com/news/grade-point/wp/2016/01/27/if-you-want-to-carry-a-gun-on-campus-these-states-say-yes/?utm_term=.f2fc364615c9

James H. Price, Amy Thompson, Jagdish Khubchandani, Joseph Dake, Erica Payton, and Karen Teeple, "University Presidents' Perceptions and Practice Regarding the Carrying of Concealed Handguns on College Campuses," *Journal of American College Health*, Vol. 62, Issue 7, 2014.

Larry Buchanan, Josh Keller, Richard A. Oppel Jr. and Daniel Victor, "How They Got Their Guns," *New York Times*, June 12, 2016, https://www.nytimes.com/interactive/2015/10/03/us/how-mass-shooters-got-their-guns.html

Tom McKay, "This Is How Chicago Gets Flooded With Illegal Guns," *Mic*, November 3, 2015, https://mic.com/articles/127842/this-is-how-chicago-gets-flooded-with-illegal-guns#.8cKjwwj75

Heather Dockray, "Igor Volsky shames politicians who send 'prayers' and accept NRA contributions," *Mashable*, June 13, 2016, http://mashable.com/2016/06/13/igor-volsky-politicians-donations-nra/#2uyBUU4CXmq2

thirty, or Come Here Fear

20

New city, new neighborhood, new apartment: a three-bedroom in the Ukrainian Village at Ashland and Augusta. We'd been there a couple of months when we came home to find our front door lifted off its hinges and leaning in the outside hallway, our rooms stripped near bare.

20

New city, new neighborhood, new apartment: a two-bedroom in Wicker Park, the first floor of a three-story crumbling Victorian. It had a fireplace, stained glass windows, and in 1995 cost $650 a month split three ways. (In 2016, that's probably tripled, if not quadrupled.) It was around the corner from Quimby's back when it was on Damen, and Myopic Books was on Division, and Earwax on Milwaukee rented art house films upstairs. I logged nights at Urbis Orbis, drinking too much coffee and reading Anaïs Nin. After I got a fake ID, I went to Double Door and wished I was a rock star. I went to Rainbo, which felt like my very own secret and was the first place I tasted whiskey.

I went to Innertown Pub and watched my roommate Laura play pool (the boys were so pissed; she kicked all their asses). Everyone was artsy and grungy and broke as hell, working and going to school, working and making art, working to keep their rehearsal space or their studio in the Flatiron, working tons of jobs, everyone from working families, families fighting for their homes as the neighborhood changed and changed and changed. When the Starbucks moved in at the intersection of Milwaukee, Damen, and North—people will tell you it's called Six Corners, but trust me, it's the Crotch—somebody kept throwing bricks through the windows. They'd get taped up, then replaced, then bricked again because: *Capitalism!* And *Corporatization!* And *Don't gentrify our neighborhood!* even though the neighborhood was already gentrifying, was already gentri*fied*, and at twenty I was too young and too dumb to understand that I was part of the problem.

But before all that, Jeff walked me home. He was my neighbor, kitty-corner across the park. We had a late-night screenwriting class together in the South Loop and afterward, we'd take the Blue Line home to Damen. He was tall, blond, small-town midwestern like me, but had just returned from living in Spain. He spoke Spanish fluently, his *c*'s like *th*'s. He introduced me to Garcia Lorca, Pedro Almodovar, and musicians that set me on fire: Heroes del Silencio, Alejandro Sanz, and Rosana.

Later, he'd read my nervous starts at stories and ask questions that gave me ideas.

Later, we'd spend afternoons at the Esquire or the Landmark watching wonderful independent films about horribly depressing people and awful blockbusters about aliens or vampires or white men in tuxedos.

Later, he'd tell me what he really thought about the guys I dated: *Your boyfriend looks like a troll doll, you know, with the hair? Wasn't that your boyfriend passed out on the floor at Swank Frank? Your boyfriend's Vonnegut tattoo is like totally original.*

Later we'd hold hands helpless watching the movie *Amélie* at the Music Box Theatre while the United States dropped bombs on Iraq, stuck in our fear and hope and privilege.

Later, I'd read the final draft of his first novel before he turned it in to Simon and Schuster, knowing the years of blood/sweat/tears that went into those pages.

Later, my boyfriend and I would move in with him when we came back from Prague, broke and culture shocked with nowhere else to go.

Later, I'd type for him as his broken hands healed from a car accident.

Later, he'd hold my newborn son while I locked myself in the bathroom and pretended everything was fine, and later—a lifetime later—he'd read this and cry.

But before any of that happened, he was the first boy I ever asked out.

I was so scared. What do we do if they say no? How do we even keep living?

"I'd love to have dinner with you!" he said. "You know I'm gay, right?"

20

I took a class on structural parody with the writer Patricia Ann McNair. It's where I learned how to break down a text, and I've returned to those lessons again and again in my professional life: the art that I make, the work that pays the rent, and the rare magic moments when those two things collide. One of the stories we dug into was Kafka's "In the Penal Colony," now one of my favorites, but then—

"I HATE THIS," I said in class.

Patty is infinitely patient but takes zero shit. "Why?" she asked.

I went off, ending with the typically exasperated: "It doesn't make sense!"

Patty nodded. She set her book on the floor. Then she leaned forward and said the single most important thing I heard in college, if not ever: "You don't get to hate something just because you don't understand it."

She let the words sink in.

Let the words sink in.

Then she picked her book off the floor and tossed it in my lap: Kafka's *The Complete Stories*. "Read it again," she said.

Every week, I stayed after class and we talked about that story: craft, historical context, its relevance to the current cultural dialogue. Every time I read it I noticed something different. "Great," Patty would say. "Read it again."

On the last day of class, she asked me to stay after one more time. "Listen," she said. She set her book on the floor. Then she leaned forward and said the second most important thing I heard in college: "Have you ever thought about teaching?"

20

The first time I gave a reading, I shook like hell. There were twenty-plus other writing students at this open mic, a very safe and encouraging environment, and I had a great story about a cocaine addict who worked in a pizza parlor and mixed up his coke with the flour and all his customers came back night after night. I got up to the podium with my two double-spaced pages and made it halfway through the first sentence before I started to shake. Like, hard. The papers rattled. I set them down and gripped both sides of the podium to steady myself, but I was still shaking so the whole podium shook—it was banging into the ground—but still, I kept going. I put both hands behind my back and grasped the elbow of the opposite arm. The shaking didn't stop, but at least it wasn't as noticeable? Maybe? I have the kind of pale (read: translucent) skin that gets blotchy when I'm embarrassed (or excited or pissed or crying or drinking or making out or thinking or breathing), so I was super red. It was obvious and horrible and, in retrospect, one of the greatest things I've ever done. Where would any of us be if we hadn't started somewhere?

20

I read somewhere that there's enough power in the female orgasm to light an entire city block. That can be a little scary the first time.

20

During the 1995–96 NBA season, I waited tables in an Italian restaurant on Van Buren not far from the United Center. That year, the Bulls beat the Miami Heat. They beat the New York Knicks. They beat the Orlando Magic and the Seattle SuperSonics and before every winning game, every step closer to the second three-peat and securing their position as one of the top ten teams in NBA history—people ate.

One night, I went into the kitchen and the chef was cutting up a goat, its carcass spread across the workstation. I don't remember how it got started, but somebody bet me a hundred bucks to swallow the eyeballs. I wasn't grossed out by dead things. They'd been in the garage my entire life. Plus, I needed the cash. That weekend I did acid and couldn't shake the idea of eyes in my stomach. Maybe they were tracking my whereabouts for government operatives. Maybe they'd grow into something terrifying that would burst out of my body like in the *Alien* movies. I don't remember who I was with that night, but whoever she was, I sobbed in her lap about my love for Sigourney Weaver.

21

A guy took me home to his meet his parents; cross-country drive, family wedding, whole nine yards. I was terrified. What if they didn't like me?

Turns out they liked me more than he did.

21

New neighborhood, new apartment: a two-bedroom in Logan Square. I didn't spend much time there. I was at school. I was at work. I was at the Empty Bottle, the Subterranean, Estelle's, Rainbo. One night I saw an ex's new girlfriend in the bathroom at Metro. She was wearing sunglasses, and she took them off at the mirror to fix her makeup.

Both of her eyes were black.

It didn't happen to me. But it happened to someone. It's happening to someone all the time. Right now, as you're reading this—it's happening.

21
My dad left a message on my voice mail: "Hey kid. I quit my job."

21
My dad left a message on my voice mail: "Hey, kid. I'm moving to Alaska."

22
My dad left a message on my voice mail: "Hey, kid. I'm getting married."

22
"I'm scared of writing."

I said this to Randy, my favorite teacher. Now I wonder what exactly I meant. Scared of the stories I had to tell? Scared of what I'd discover? How young I was, how naive? Scared of what my parents would think? Scared of what reviewers would think? Scared I wasn't good enough, whatever that meant? Scared of trying to make a living as a writer, sans 401K, eating cat food? Scared that the things that mattered to me would seem trivial to others, and if not trivial, then cruel or stupid or wrong?

Randy listened. He is what listening looks like: leaning forward and fully engaged, whether he's talking to Salman Rushdie, an interim provost, my seven-year-old son, or a twenty-two-year-old student figuring out what the hell she was doing. "Fear is a logical response," he said. "This is your life." We were in his office, high above Michigan Avenue. One windowed wall looked over the lake; everything else was books. "But—" he said, and I've returned to this moment a thousand times, when writing is too the fuck much, when higher ed administration gets in the way of higher ed's mission, and when I talk to my own students about their hopes and fears—"you're having fun, right?"

He turned to his books, searching stacks till he found the one he wanted, searching pages till he found the magic that would fix my whole life. He's done this many times over the past two decades, but on that particular day it was the final passage of Kafka's *Diaries:*

> More and more fearful as I write. It is understandable. Every word, twisted in the hands of the spirits—this twist of the hand is their characteristic gesture—becomes a spear turned against the speaker. Most especially a remark like this. And so ad infinitum. The only consolation would be: it happens whether you like or no. And what you like is of infinitesimally little help. More consolation is this: You too have your tools.

The last line was underlined.

"You too have your tools."

I don't know what that meant for Kafka. I don't know what it means for you, but for me it's the books and stories and poems and movies and art and songs that save us.

22

Back then, the creative writing department where I finished my undergraduate degree offered a combined MFA in writing and the teaching of writing. I thought that was a way I might be of use: show others how to use their voices the way my teachers showed me. I got some fellowships, and my mom helped, but for the most part I worked at the Bongo Room after it moved from Damen to Milwaukee Avenue. I was there for over a decade. The lines for brunch still stretch down the street. Everyone in Chicago knows those pancakes: chocolate mascarpone and pumpkin spice and lemon blueberry whipped cream. More than one customer asked me if the secret ingredient was cocaine. *Shhhhhh*, I'd say.

At one point I started counting the pancakes I carried across the room. How many equal a graduate degree?

I'll tell you what: that program was worth every fucking second.

I sat in semicircles with incredible writers. They made me work harder. I'd come home exhausted wanting to drink beer and watch *The Simpsons*, but then I'd think of that semicircle. They were awake, I knew: writing, reading, working, digging deep, and I'd turn off the TV and show up at the computer. Here I am! I'd announce to no one. I! Am! Here!

The one who pushed me the hardest was a guy from Kentucky named Lott. One weekend a month, we'd drive through Indiana to the Michigan side of Lake Michigan and write at his friend's beach house, which had once been owned by Carl Sandburg. It had five bedrooms, four stories, with a tiny elevator, and a porch on the roof with rickety stairs leading to a widow's peak. Winter lake winds would pound against the windows. Lott would make fires in the fireplace and we'd play made-up drinking games, writing our fears on little scraps of paper, putting them in a hat, and telling stories one by one. Love. Loneliness. Thesis (the word "thesis" was always followed by the word "fuck": *thesisfuck).* "Have you written about that yet?" he'd ask, and when I said no he'd give me a look.

Later, he'd read my nervous starts at stories and ask questions that gave me ideas.

Later, on the way home from a Halloween party, some guys in a passing car would throw eggs at me and he'd run after them down Armitage Avenue and some drunk frat boys dressed as nuns would see him and yell: "What are you doing?" And he yelled back: "That car egged my friend!" And then the nuns ran down Armitage with him and they caught up with the car at a red light and had a little chat about manners.

Later, he'd tell me—so carefully, so gently—that my boyfriend was actually gay.

Later, we'd co-teach college classes on art and community engagement.

Later, he'd get ordained on the Internet to marry my husband and me on the beach at Lake Macatawa, the most perfect day in my life.

Later, when my son was six months old and puking all over the house, he and his husband, Ryan, came over so my husband and I could go to the hospital because we'd started puking, and then they started puking, and when I tried to apologize, he sighed and said, "We're family. We're supposed to infect each other."

Later, he was the executive director at the faculty development center where I worked for the better part of a decade.

Later, when he was told to fire me, we'd get stupid drunk and I'd say, "Do it."

And he'd say, "I can't."

And I'd say, "You have to."

And he'd say, "Okay. Okay. Okay—you're fired," and we'd laugh so hard we knocked over our prosecco.

Later he and Ryan would take my son for weekends all summer so I could write this book.

Later they'd send us pictures from their beautiful new home in California and I'd cry because I was happy for them and I missed them and where do you put all that love?

But before any of that happened, we sat in front of the fire at Carl Sandburg's house on the Michigan side of Lake Michigan.

"Have you written about that yet?" he asked, and gave me a look.

It meant: *It's time.*

22

I was out to hear a band and a guy I didn't know put his hand on my ass. The place was packed, everyone body to body, but this was no accidental brushup, no quick cop a feel. He reached around my hip, grabbed a fistful of my ass, pulled me into him, and grinned. "Let go," I said, trying to jerk back, but he was too—I don't want to say *strong*. Using your power, be it physical, societal, or financial to intimidate or manipulate or take something not given—that's not strength. This guy was bigger than me. He was *not* stronger. "Let go," I said again, but he yanked me closer, pressing his crotch into my stomach so I'd know his cock was hard and said, still with that pretentious fucking grin,

"You have to say please."

I'd like to tell you that I spat in his face or kneed him in the balls or staked him through the heart, but I can't. "At core, men are afraid women will laugh at them," wrote Gavin de Becker in *The Gift of Fear*. "While at core, women are afraid men will kill them." I hate admitting that. I hate knowing how fast and often these moments turn violent. I hate that afterward I asked the bouncer to walk me to my car.* I hate how I still don't listen to that band. I hate the memories that show up uninvited every time I go out dancing, or park my car on a side street, or swim, or clean the floor, or walk into the stairwell at the college where I worked or the restaurant where I worked or the L or the street between the L and my apartment or the alley where I walk my dog every morning or any of a thousand places. There are a thousand stories. Here's why I chose this one:

Please.

I said it. I said please. I asked this asshole, nicely, to give me back my body. Look: I know how small this experience is in comparison, but whenever I hear talk about the "right" way to stand up for yourself, the "right" way to protest, I taste that please: bitter, burning, furious. Ask nicely to make decisions about your own body. Ask nicely for police to stop killing you. Ask nicely for your family to not be deported. Be patient as we discuss whether or not you may go to the bathroom. Be patient as we decide whether or not you are allowed to marry the person you're in love with. Be calm when there's a gun in your face. Be

---

* I'd like to thank that bouncer. I know it's not necessary; he was doing his job, doing the right thing, and taking care of each other should be the default. But as I look back down the line of my life at the moments when I was the most afraid or the most alone or closer to the ledge than any of us are safe to be, it was always kindness that brought me back.

calm when there are tanks on your street. Be polite in the tear gas. Smile as your schools are closed. Don't be so fucking sensitive.

23

I don't remember his name. I don't remember what he looked like. But he was the first man who went down on me when I was on my period, who wasn't afraid of blood and more so liked it, loved it, looked up from between my thighs with his face smeared red and I came so hard I slammed the back of my head against the wall.

23

On one of our last nights in the apartment on Armitage, I tried to make dinner for Heather and Pete. This was a big deal. I don't cook. Left to my own devices, I would subsist off of pinot noir and a cheese plate.

Half of the apartment was already packed, books and clothes and duct tape everywhere, but the kitchen was still untouched. Truly, it was a biohazard. I'm surprised we're still alive. Heather and Pete had gone to Jewel-Osco for more boxes and I made tuna casserole because hey, Who can fuck up tuna casserole? Noodles, cheese, peas—voilà. I put it in the oven and went to work on the mess stacked in the sink, plates and pots and mugs and flatware. It's possible they'd been there since the day we moved in. We had zero counter space, so I got two bath towels and laid them on the floor, putting dirty stuff on one so I could get at the sink and newly clean stuff on the other to dry. Some of you reading this are disgusted with us. Some of you have been there. Some are there right now. I'm here to tell you to forget about the dishes. Fuck the dishes. Grab your friends and go dancing: hands here, feet here, ass like this. Grab them and tell stories: "Poor baby. You just want to feel all right." Grab them and hold on. It may be the last time you're all together.

I burnt the casserole, but you knew that was coming. I also forgot to add the tuna; maybe you knew that, too. Pete ordered a pizza and we went into his room, away from the burnt smell, and ate it on the floor. I remember how empty it was with everything in boxes. I remember the shock of the walls, recently painted a neutral color per our agreement with the landlord. The angry, furious paintings on grocery bags had been removed long before. I don't know how long he kept them up.

A day? A week?

In my memory, they're still there.

24

New neighborhood, new apartment: a broken-down, termite-infested fucking awesome yellow two-flat at the corner of North and Humboldt, a block from the park. There were mint bushes everywhere. So much mint. The mint went insane like the tomatoes in *The Witches of Eastwick*. It overtook the whole yard, the whole house; we drank mojitos 24/7. We drank wine and whiskey and too much coffee and ate leftovers from fancy restaurants where friends worked and cheap pizza delivered at 2:00 a.m. Like in many Chicago apartments, the bedrooms were tiny; once you got the futon inside you couldn't open the door all the way. Stay off the back porch—it's a death trap. In the summers, electric bills skyrocket from portable air conditioners and in the winter we froze, even after meticulously insulating the windows with Saran Wrap–like shrink film sealed with hair dryers.

I lived upstairs with my friend Sue, and Mike and Dia were in the apartment downstairs. In the mornings I'd drink all their coffee while Dia finished her thesis, an enormous quilt made from clothing given to her by different women along with the story of what it meant to them. I gave her one of the nylon grandmother slips I'd worn on Martha's Vineyard, telling her how sexy, how powerful, how free it made me feel. We were piloting an arts integration program called AIM, teaching writing through textile art in elementary schools around Chicago. We wrote curriculum and waited tables and thought we could change the world.

Later, she'd read my nervous starts at stories and ask questions that gave me ideas.

Later, she'd sit at her sewing machine in our basement/fashion studio/craft studio/recording studio/dance club/rehearsal space/smoking lounge and make clothes that actually fit my body.

Later, when I was in the guts of a particularly gnarly breakup, she'd lock me in her bedroom and play Ben Harper's "*Walk Away*" on repeat: "Oh no/ Here comes the sun again/ That means another day/ Without you my friend."

Later, she'd move to San Francisco and I couldn't stop crying, the sad kind.

Later, I'd go to San Francisco to meet her girlfriend Jessica and I couldn't stop crying, the happy kind.

Later, she'd help me put on my wedding dress because I kept getting it stuck in the Spanx and she was like FUCK SPANX and I was like YOU'RE RIGHT FUCK SPANX and I didn't wear the fucking Spanx.

Later, she'd call to say I couldn't help her put on *her* wedding dress because she and Jess had to get married right then, that day, before Prop 8.

Later, she'd hold my newborn son while I locked myself in the bathroom and pretended everything was fine.

Later I'd hold *her* newborn son while my three-year-old pet his tiny face and said, "Cousin, cousin, cousin!"

Later we'd hold hands helpless watching multiple real-time Twitter feeds broadcast Michael Brown's murder and the protests in Ferguson: riot gear, tear gas, tanks. I went to check on our boys, their little bodies wrapped together in sleep, and I understood something then like *lightbulb, lightning bolt, ton of bricks*: as a black mother, she has to talk to her black son about how to walk in this world and not be harmed, and as a white mother, I have to talk to my white son about how to walk in this world and not perpetuate that harm, to stand up when we see it and fight it like a dragon.

But before any of that happened, it was the four of us together in that yellow two-flat, my first understanding of chosen family.

If I was scared of anything, it was knowing it would end.

24

EXT. BACK PORCH—DAY

Megan and Dia sit in lawn chairs. It's the last day of 1999.

MEGAN:

Computers will reset to the nineteen hundreds. Everyone will panic, exposed and off guard. That's when they'll strike.

It's like Skynet. It is Skynet. They came back from the future and planted the *Terminator* movies so we'd have context for the takeover. You know sci-fi is real, right? I mean, all fiction is real, but sci-fi is like, prophetic. H. G. Wells wrote about the atomic bomb in nineteen fourteen. E. M. Forster wrote about cubicles—cubicles!—in nineteen hundred and nine. Before Oculus Rift porn there was *Demolition Man*, and escalators and debit cards and face-scan technology and sure, fine, some of that stuff is cool as hell but for the most part, we're fucked.

DIA:

How much pot have you smoked today?

MEGAN:

(starts to cry)
Tonight we all die.

DIA:

Who cares so long as our student loans are wiped.

24

I graduated with my MFA in May of 2000. The ceremony took place in Chicago at Navy Pier, a long stretch of shops and restaurants and theaters sticking out of the Loop and into Lake Michigan, all neon color and swarms of people milling about. My grandparents and my mom had driven in from Michigan to support me and to take a million pictures.

I walked across the outdoor stage at Navy Pier, hugging Randy and receiving my diploma. The whole process took ten seconds. It was freezing cold in May. I sat with my friends Lott and Joe and shivered. The commencement speaker went on and on. I don't remember about what. I wanted to get out of my seat and ask my family if we should sneak out and grab a beer. But for the ten seconds I was up there, I looked out into the crowd and I saw my grandpa. My mom and grandma and friends were applauding and taking pictures, but my grandpa just stood there with this smile on his face. I knew how proud he was of those ten seconds, and the seven years it had taken me to get to those ten seconds, and I knew we'd both sit in that freezing outdoor auditorium for the next three days if need be: he'd sit there for me, and I'd sit there for him.

In the center of Navy Pier is a Ferris wheel. It's big, way up high in the sky. It takes an hour to ride the long, slow circle through the air and back around to the ground. My grandfather stood under that Ferris wheel and stared up. Then he looked at me. I don't do heights but I did that day. We didn't talk much, which was rare 'cause my grandpa and I could talk. We talked about the war and the stock market and his family and my future. We argued a lot. Once, when I was in high school, we had a real intense one about some charged political issue I don't even remember now and I left his house furious. A few weeks later he sent me a letter about how he'd been thinking about what I'd said. Had talked to some friends about it, and the priest at his church. And I think that's the most important thing I ever learned from my grandfather: No matter how set in our ways, we still have much to learn. We can listen. We can try. That *is* possible.

25

I slept with a guy who made me keep my socks on. He was afraid of feet.

26

On days I worked at the Bongo Room, my alarm went off at 6:00 a.m. to get down North Avenue and open the restaurant by 7:00. On my days off, I slept in till 8:00 a.m.—9:00 eastern standard time—and on September 11 I walked into the living room just in time to see the second plane crash into the South Tower. I didn't know what I was seeing. Sue would get home late from the bar and fall asleep on the couch watching movies. I assumed she'd left the TV on. I peed, made coffee, and came back into the living room. The towers were still on the screen, both smoking, the commentators' voices panicked and unsure. I went to Sue's bedroom; she was already gone for an early morning class. I went downstairs; Dia and Mike were awake and on the phone. I called Jeff and he came over. Sue came home with her boyfriend. We sat together on the couch, and on the floor around the couch, everyone's body touching everybody else's body so we knew we were all there, that no one was alone. We stared at the news for hours. For weeks. For years. I think we're all still staring.

Now I get the majority of my news from social media, staring at screens, moving between different feeds, more perspectives and greater truths. I miss having someone next to me, a leg pressed into mine, shoulders locked, body to body and together, breathing, living, alive.

26

A student leaves a note in my faculty mailbox, one of those pink slips saying IMPORTANT MESSAGE at the top. It's signed by her psychiatrist and says she won't be coming to class. She's afraid of anthrax.

Anthrax is all over the news—"one part Clorox to nine parts water"—but until that very second I hadn't worried about its proximity to me, probably because I'd been busy worrying about other things: friends in New York; the racial profiling and violence experienced by Arab American friends and students and human beings everywhere; the Coast Guard base where my dad worked that was on lockdown and I couldn't get in touch with him; the families who didn't know what happened to their loved ones and those five hours I sat in Boston, watching the news, not knowing, not knowing, not knowing; people worldwide who experience this sort of violence and worse on a day-to-day basis and my own privilege as an American; something I read about the Sears Tower being next; something I read about airplanes as time bombs; something I read about reinstating the draft; and how am I supposed to teach writing? There we were, eighteen—no, seventeen—college seniors in a semicircle around me and I'm holding a piece of paper about anthrax and looking up at the vents in the ceiling imagining white fluffy gas pouring into the room and me yelling hit the deck and all of us jumping on the floor on our stomachs, wrapping our hands over our noses and mouths, and finally someone said, "Megan?"

My mouth was full of gum. I could reach in and pull it out, a huge pink pile at my side. Everyone sat there, waiting for me to say something, and finally I looked up and asked, "Are you afraid?"

One young woman said, "Yes. Airplanes. I'm afraid to fly." And everybody nodded. I went to the board and wrote "airplanes."

"What else?" I said, and we started a list: bombs, racial profiling, biological weapons, domestic terrorism, student loans, not getting a job after college, invasion of privacy, the draft, corporate media, white people, Republicans—

"I don't want to think about this," someone said.

"That's the problem!" said someone else. "Nobody's thinking about their fear, no one's deconstructing it, so we're all irrational and terrified and dangerous and stupid and fucked."

I thought about how unequipped I was to have this conversation. I was such a new teacher, so young, so green, and at most institutions, training of any kind—instructional design, crisis management, diversity and inclusivity, unconscious bias, college-wide resources, sexual assault, educational theory, educational technology, you name it—is nil. *Here's your classroom, here's a roster, go get 'em, tiger.*

"Look," I said. "I'm not sure what I'm doing. Not in my writing, my life, and right here, right now. You don't graduate from college and immediately it's voilà! Here are the answers! You find them as you go. You make mistakes and fuck up and fix it the next time around, and maybe it's a huge mistake for me to be talking about fear so soon but it's all I can think of right now, so"—* I took a deep breath—"I'm afraid of

---

* Years later, reading *Teaching to Transgress* by bell hooks, I would find this passage, unwrapping it like a gift:

"In my classrooms, I do not expect students to take any risks that I

crowds," I said. And I told them about having a panic attack at a George Clinton concert.

"You should have known better," somebody said. "There's like a million people at those things."

And somebody else said, "But I'm afraid of elevators, and you can't, like, not ride an elevator."

And somebody else said, "D'uh, stairs."

And somebody else said, "Not when you're late for class and it's on the twelfth floor and—"

"Hey!" I said, writing elevators on the board. "What else?"

Germs, bugs, tap water, chewing gum—some people laughed. So we paused to set the collective expectation that we were entering this brainstorming session with curiosity, not judgment.

Fire, heights, needles, doctors, ghosts, rain, sex, sexual assault, guns, Conceal Carry legislation, conversion therapy, Republicans—"We already said that!"

Big dogs, little dogs—that last one was mine.

Criticism—we stopped again to break that one down: constructive criticism versus being an asshole.

Oil paint: "It can kill you!"

Feathers: "So many germs."

Water: "So many germs."

Mice: "OMG, germs!"

Carbonated beverages, camping, cancer, cancer, cancer, cancer,

---

would not take, to share in any way I would not share. When professors bring narratives of their experiences into classroom discussions it eliminates the possibility that we can function as all-knowing, silent interrogators. It is often productive if professors take the first risk, linking confessional narratives to academic discussions so as to show how experience can illuminate and enhance our understanding of academic materials. But most professors must practice being vulnerable in the classroom, being wholly present in mind, body, and spirit."

*cancer, cancer, The Ring, glue, blood, the dark, the police, failure, parents' death, child's death, being alone, dying alone, dying period, OD'ing, God, organized religion, paralysis, marrying your second choice, zombies, not being heard, not being seen.**

26

I got dumped. It was tragic. I loved him so much. I'd have loved him forever. He didn't love me back and I thought I might die.

I didn't want to go home, so I went—again, always—to Dia's bar, waiting for her to close. She kept refilling my drink, so I'm sitting there drinking and crying and drunk and ridiculous, mascara all over my face, snot everywhere, and I hear, "Megan?" I turn around. It's one of my college students. It was so awful. It was so awkward. He didn't know what to say, poor guy, as if he could've said anything in that moment, and in the end he was like: "But you're supposed to have your shit together!" To him, I was only his teacher: four hours, once a week, knowledgeable, professional, shit together. And that's true. I am that person. In classrooms, conferences, festivals, meetings, retreats, workshops, presentations, performances.

---

* I copied this list after class and found it, years later, while working on this book. Hence:

Dear Present and Future Personal Essayists (and writers of all genres and artists of all genres and human beings, period): Be specific in your journal. Names, places, weather, dialogue, sounds, smells. Later you'll be grateful you had such foresight. I pulled extensively from past journals for these essays, and there were many, many times that forty-year-old me had no idea what twenty-something me was referring to, especially in the passages about love. I'd be all: I LOVE THIS PERSON SO MUUUUUCH I'LL LOVE THEM FOREEEEEEVER THEY DON'T LOVE ME BAAAAAAAACK I MIGHT DIIIIIIEEEEEEE.

Now I'm like, who the hell was I talking about?

But there are other parts of me, then and still, that are equally true. I want to sit on the couch and watch *Orphan Black* and not think about anything. I want to sit with my husband and a bottle of wine and not have to be *on*. I want to sit on the floor and play with my kid. I want to sit at my laptop and read the news and cry because so much is awful and it's so hard to stay hopeful and on those days, I shouldn't be anybody's cheerleader or mentor or teacher. I don't know how to help you. I don't want to read your manuscript. No you can't pick my brain over coffee. I don't have time for coffee. I shouldn't drink this much coffee, I shouldn't eat this bagel, I want this fucking bagel, fuck you, bagel, fuck you, carbs, fuck you, thyroid. I say fuck a lot. I say the wrong thing. I say the right thing the wrong way. I don't want to say anything at all. I'm tired. I'm tired. I'm tired.

26

Same neighborhood: new one-bedroom apartment a block down the street. Sue moved to Florida. Dia moved to San Francisco.

It was the only time I'd ever lived alone.

I know some people love it.

But if I was scared of anything, it was knowing this could last.

26

The first time I performed a story live was at Schubas, a rock club. We opened for EXO, a rock band. I was onstage with Julie, a rock star, and I was terrified. My arms were shaking. The pages in my hand were shaking. If I'd had a podium, it would've shook, too. I remember standing next to the little stairs leading up to the stage thinking how dumb this was. I was supposed to be locked in my room with a typewriter and a bottle of Wild Turkey.

That's when EXO's drummer, Doug, came up to me. He was super tall with shaggy hair, and every time I saw him, he had this look on his face like he'd just heard something hilarious.

"It's time," he said, nodding at the stage where Julie and Robert, our guitarist, were setting up.

"I don't think I can," I said. And I will never forget this. He said, "Megan. This is the fun part."

26

My grandpa tells me to get a full-time job. This adjunct teaching business is a racket. If I insist on writing, I can do so on the weekends. "You have a masters degree," he yells. "You should not be waiting tables!"

"The smartest people I know wait tables!" I yell back.

He pauses, confused. "Why are they doing that?"

I explain that I make more money pouring mimosas than I do teaching college students.

Let's sit quietly for a moment and consider what this says about our culture.

"That's awful!" my grandpa yells.

"Agreed!" I yell.

"But you need health insurance! A 401K!"

"I have health insurance! And an IRA!"

"But—"

"And I love my job! I love all of my jobs! How many people truly love the work they do?"

"But—"

"And I have time to write—sort of."

"Megan—"

"And I'm helping people! I think? Maybe? I'm trying!"

My grandpa throws his hands in the air. He's dramatic like that. Dramatic like me. "Fine," he says.

"Fine," I say. "I win!"

"You had to win," he tells me. "You can't lose a fight about your own happiness. You can't lose a fight about your own life."

26

A guy I loved sank all his money into an RV and asked if I wanted to drive across the country. I wasn't teaching summer school. And covering two months of brunch shifts was doable, as was a temporary sublet on my apartment. I said sure. I knew I was setting myself up—he wanted adventure and I wanted him—but I decided to take the odds.

That's what life's about, right? *Risk*.

You go over the waterfall. It's what the waterfall is for!

You drive across the country in an RV. It's what an RV is for!

"Give me a sign!" I said to the sky, spinning in the courtyard. It was May. May in Chicago is perfection. We made it through a shitty winter and anything is possible.

On his way to pick me up from Michigan, the RV caught on fire.

27/28

I'm standing on Petrin Hill overlooking nighttime Prague, lit up like a miracle. My students brought me here for my birthday: an uphill funicular ride to Restaurant Nebozizek, a glass conservatory in the trees. It had a live piano player.

It had delicious food. It had champagne—a lot of it—and at the end of the night, I took my glass on the patio. It was August, warm winded and lovely. The stars in a grid. The city in panorama. I was so desperately in love with this new country, this new life. We'd been there a few months, twenty students and I plus two other teachers, reading and writing and getting lost in the castles and cobblestone. Randy'd needed to staff a Kafka class for the department's study abroad program and knew I was obsessed/infuriated/hypnotized/a hot mess about the man and his work—in academia, this is referred to as *a scholar*—and was I perhaps available? Six months later, my students and I were in the back room of Café Montmartre, a tiny coffeehouse off Retezova where Kafka read aloud to his friends. We dug into his writing, his life. We felt him in the walls. We cased his city: his birthplace near St. Nicholas; the Workers Accident Insurance Institute now a hotel; the memorial on Dusni Street designed by Jaroslav Rona (it's weird as hell and totally perfect); his grave at the New Jewish Cemetery; and, my favorite, the blue house at number Twenty-Two, Zlata Ulicka, where he wrote stories for *A Country Doctor* and, later, *The Castle.* I stood in that tiny, near claustrophobic room and let myself dream.

That night on Petrin Hill, I came up with a plan: I'd go back to Chicago, back to the Bongo Room and ask for my job. I'd live off what I made teaching and save my tips. Then, after teaching next summer's study abroad session, I'd stay in Prague and write until my money ran out.

Later, I took the funicular back down to the city and got a tattoo to remind me of that decision, the stars in a grid on my inner right biceps.

Listen: never get a tattoo when you're drunk.

28

A week later at Café Montmartre during Kafka class, my students and I arrived at the end of *Diaries*—"more and more fearful as I write"—with the power punch of a last sentence that had so long been my lifeline:

"You too have your tools."

But by the end of the paragraph I realized we had a different translation.

It didn't say tools.

It said weapons.

28

Because I'll be moving to Prague at the end of the year, I decide that dating is a waste of time. Why put in all the effort of building a relationship? I have my friends and if I need sex, I'll go have sex! That's a thing you can just do, right?* Just go out there and have sex?†

But then something crazy happened.

28

I fell in love.

28

New country, new neighborhood, new flat: from Wenceslas Square, take the green line metro to Náměstí Míru. You'll see

* Yes. Have fun!

† Yes. Be safe!

the Church of St. Ludmila, an enormous neo-Gothic castle-like place with a wide courtyard. Cross the tram line and hang a left onto Belgická. A half block on the right is an Internet café—40 Kč for an hour, which in 2004 was a little over a dollar. Another block forward and the streets change from asphalt to cobblestone. On the right is Medůza's, the café where I wrote every morning and on the next corner, a Mexican place called Žlutá pumpa with raspberry margaritas. A half block from there, on the opposite side of the street—see the pink building? The top floor right was ours.

I liked that word.

Ours.

There, Christopher and I built a relationship. If he said or did something that I didn't understand, I couldn't go out for margaritas with my girlfriends and say, "What does that mean?" I had to ask him.

There's so much bullshit in our beginnings, so much baggage and fear.

For better or worse, we faced each other.

29

I had more sex than I've ever had in my life.

29

I wrote more than I ever had in my life.

29

I remembered how to start thunderstorms with my brain.

29

We had time to do the things you never have time for. Slow cooking paella, daylong getting lost in the winding streets around Old Town Square, reading a whole book in a single day, watching all four *Alien* movies back-to-back. Christopher, for reasons that now elude us both, decided we would watch all the war movies: *Empire of the Sun, Schindler's List, Full Metal Jacket, Saving Private Ryan, The Bridge on the River Kwai, Good Morning, Vietnam, The Thin Red Line.* For weeks, our flat was thick with aggression and despair. I cracked on *Platoon*, rolled up on the bathroom floor and unable to stop crying.

Here is my privilege: I could turn away.

29

Our friend Marketa told us: "It is okay. George Bush is not your fault. The world knows he stole the election. It will be fine in November."

29

Alitalia airlines had some sort of financial problem and offered ridiculous deals. We're talkin' thirty dollars roundtrip to cross the continent. We went to London, where I had a panic attack on the tube; Paris, which we forgot to leave (i.e., we went to the airport and discovered our tickets were for three days earlier); and Italy. I wanted to remember Florence, to show Christopher its museums, to walk into those stores on Via Roma and buy dresses for my nineteen-year-old self. I promised her. I promised.

Those stores, of course, were Gucci, Miu Miu, Stefanel.

Ten more years, I decided. For sure I'll have money by then! That's when you have money, right? Thirty-nine?

29

That September, a group of Chechen terrorists took more than a thousand little kids hostage, holding them for three days before Russian forces stormed the school. This event received twenty-four-hour nonstop coverage from the entire international news media. Christopher and I sat in front of our TV, watching horrified as the Russian government deliberated and the eventual, horrible climax of hundreds of children running to safety.

At one point I called my mother back in Michigan. "Aren't you watching CNN?" I asked, trying to get a hold of my words 'cause I was crying so hard.

"What are you talking about?" she said. "They're interviewing Bill Clinton about his new book."

That's when I understood how naive I was. That I hadn't considered the difference between domestic and international new sources. What gets covered depends on where you are, and who's writing. And who's editing. Proximity, nationality, economics, and politics are all lenses used to frame the truth. Another example: the media I'd consumed since leaving the United States was so fervently anti-Bush that the possibility of him winning the 2004 election hadn't occurred to me.

29

November 5, 2004. Marketa sends us a text message:

> Oh no! It's very bad! I would like to cry! I don't understand people who want to have so bad president! Don't be scared please I still like you and I'm sorry about your bad president! I wish you better time my darlings.

29

Back to Chicago, back to Humboldt Park: a new third-floor walk-up at North and Kimball. It had three bedrooms and an open floor plan. Before we signed the lease, the landlord asked what would happen when we broke up. Which one of us would move out? How would the other pay the rent?

"We're not going to break up," Christopher said.

She gave him a look.

"We're not!" I said.

She called me *my dear* and said, "Good luck."

Luck: we had each other. Luck: we had our friends. Luck: we still had jobs. I split my time in front of students and behind a bar, and it was then that I started to figure out the craft of oral storytelling, how the literary techniques I was teaching from Kafka and Morrison and Didion were so similar to those my customers used after a mimosa or two or five: structure, tension, exaggeration, scene. This is a necessary plot point to the story of me and the personal essay; I didn't start off writing them for the page. I wrote them to tell out loud, mostly in bars, to a hundred or so deliciously tipsy people looking for connection in this beautiful mess of a world. I signed on with a storytelling collective called 2nd Story; we are educators and directors and producers who support people of all ages, backgrounds, and skill levels in first writing their own stories, and then performing them out loud.

People spend a lifetime looking for a creative community. This is mine.

29

Fuck the shows and movies where the woman won't get out of bed on the morning of her thirtieth birthday. Fuck the expectations and the social conditioning. You do not need to be married with careers and houses and good skin and a size four and gold cards and Blahniks and poodles and babies—babies! I didn't have a dentist.

"I want a baby," a friend told me. "I'm thirty-four."

"Madonna—" I started.

"DON'T TELL ME MADONNA HAD A BABY AT FORTY EVERYBODY SAYS THAT I COULD GIVE A FLYING FUCK FUCK MADONNA MY CLOCK IS TICKING."

I'd read an article in *Newsweek* about women freezing their eggs to fertilize later, if and when they decided to have a family, however they chose to do so. Apparently, they were "lining up with their credit cards and their dreams."

I liked this idea.

"Wait till you're thirty," my friend said. "Just wait."

29

We got a puppy.

30

We walked into the house. I didn't know how I would feel. I hadn't been there in over a decade, not since my parents split, and my mom moved out, and my dad got a renter and left for Alaska. There were the wood walls and floors and ceiling, the front windows overlooking our little lake. The dock was still there where that boy threw turtles for my dog to fetch. The rowboat was still there, same green oars that carried me away. Upstairs were my bookshelves, the poems on the walls painted over. The garage felt wrong in its emptiness, but the rest of the house was new: new paint, new carpet, no furniture—ours.

When Dad called from Alaska and asked me to drive to Chelsea and check on the house, I was worried it would be haunted, that I'd only remember the last few years of high school when everything got hard. Instead, I saw a refuge, a hideout, like Terabithia or the Batcave. A new renter would move in that summer, which gave Christopher and me nearly a year of weekends to escape from the city. We slept on an air mattress in front of the fireplace. We spent Saturdays reading books. We figured out what would happen next.

30

We got married on a beach in Macatawa, Michigan. Our friends Amy and Scott were kind enough to share their home, and we brought ten friends for the week. It was lovely. It was easy. We swam in the lake and shopped at the farmer's market. We watched *The Shining* four times in a row. Dia and Jessica and I went skinny-dipping. Randy read us poetry. There was Maker's Mark in bulk because apparently, while everyone was getting dolled up before the ceremony, Jeff walked into the kitchen where the caterers were setting up and yelled, "MY GOD THERE'S ONLY A HALF BOTTLE OF BOURBON!" One of them took pity on him and went to the store. Later that evening he walked me across the beach to my almost husband and as the people we love looked on, Lott—by the power invested in him by www.spiritualhumanism.org—married us as the sun went down.

It was perfect.

It still is.

I'm not saying that every day is barefoot in the sand. That's okay. I can sit here in my sweatpants watching him cook mac 'n' cheese and think: *I am wild about this person.* I think it when he talks about art or worries about money or worries about me. When he rides bikes with our son or builds robots with our son or talks to our son about what it means to be a good person. When he reads books every night before bed because he's online all day and has rules about when he can and can't be on the Internet. When he's walking toward me from a distance and, in the split second before I recognize that he's mine, I think: *Oh my god, look at that guy!* And when he sits at the computer, working, working, working for this life we lead and these dreams we have—him and me and our family and our plans.

Course of One's Life

The new provost asks all faculty and staff to submit their updated CVs to his office for review. *Ask*—his word—but it's clear that this is a requirement. "If you're unsure of the conventions of a CV," his e-mail informs us, "you can find support at the Center for Innovation in Teaching Excellence."

Within the hour my inbox is full of fear.

*

In higher education, abbreviations are an art unto themselves. Ours was C-I-T-E and pronounced "the site." After years as a freelance writer and part-time educator, this was my first full-time job: first nine to five, first trip to HR, first weekly staff meeting and white elephant holiday party and e-mails about leftover donuts in the break room and please don't clip your toenails at your desk. I had a desk! My own desk! It came with a dedicated file cabinet, an ergonomic chair and was next to a window, which those of us in cubicles know is the holy goddamn grail. It reminded me of *Scotty Got an Office Job*, this incredible web series by Scotty Iseri about the absurdity of office culture. He writes lines from Pee-wee Herman on his palms and secretly films himself gesturing at meetings. He steals a coworker's teabags, the ones marked DO NOT STEAL MY TEABAGS, and stuffs them down his pants. He performs an epic synchronized swimming routine across a sea of cubicles, and every time I crossed my own sea—four rows of six, specifically—I wished I could do the same.

Truth is I was rarely there. My team and I ran from building to building across the college's South Loop campus, facilitating workshops for departments in instructional development and educational technology. We designed curriculum, online classes, and faculty retreats. We sat on committees for pedagogy, learning outcomes, and strategic planning. We did classroom observations and one-on-one consultations. We attended student forums and took their perspectives back to the faculty, centering their needs across our programming.

Specifically, my job involved helping teachers create inclusive classroom spaces, which meant bringing together the many people engaged in the work in their separate silos, a sort of lab or think tank. "I imagine it like this," I'd say. "My creative nonfiction class is a place where students can focus on their writing. Once a week, for four hours, the rest of the world falls away. So often, we don't have that luxury as teachers: a place to talk about our work, to think about our work, a place to try new things."

I loved it.

When you have love, you can handle any amount of bullshit. Right up until you can't.

*

"CV is short for curriculum vitae. Latin—" I'd say, "for 'course of one's life.'"

Laptops are out, everyone furiously taking notes and panicked about paperwork. The room is packed. A year earlier, the college's new president and CEO told an auditorium full of faculty and staff that we were on a "value clock": at some point, our value to the college would run out and we'd be let go. The air went out of the room. Tension unrolled, an unspoken fear carried around campus and into classrooms. *Shhhh, can you hear it? Tick tock*. Since that day, attendance at the CITE's professional development workshops spiked—CVs, academic cover letters, teaching philosophies and portfolios, job searches. The provost's call for updated documents shouldn't have been cause for alarm—it's standard practice at the university level and rightly so, in my opinion—but when the whole of your job becomes defending that job, it's hard to see through the fear.

My office drowned.

Hi, Megan. I've attached my CV. Can you look it over and tell me what to do? It's no secret that higher ed is in trouble; full-time positions are few and far between, tuition is climbing, enrollment dropping, tenure under fire, and more than half of the country's college teachers are adjunct, trying to piece together a living wage. I pored over their documents, looking for places I could help. If we move this award higher, will that get your foot in the door? If you list teaching credentials on page three instead of page six, will your job be protected? If you change your second reference, will you be able to feed your kids?

*

Early into my new position, I facilitated a retreat for the film department. The morning of, I was told they'd changed their name and so, in my opening remarks to the fifty-some faculty and staff in the room, laying out the day's agenda and so forth, I welcomed the new Department of Cinema Arts and Sciences.

A hand was raised. An administrator stood. And he explained, to me and to everyone, that the department was not "Cinema Arts and Sciences" plural, but "Cinema Art and Science" singular, a decision that had been made collaboratively with full transparency following robust dialogue unpacking the profound complication of Cinema Arts and Sciences being confused with the School of Liberal Arts and Sciences who already held ownership over said plural form and additionally henceforth the "*and*" in Art and Science would be demonstrated textually not as *a-n-d* or an ampersand (&) but rather a plus sign (+) in all branding and documentation—however when spoken aloud we will of course use *a-n-d and* for a more common understanding.

I laughed out loud.

Mine is not a delicate tee-hee little golf clap of a laugh. Jaw goes down, eyebrows hit hairline and I snort. Charming as hell. You want me at your parties.

The look on this man's face said that was not an appropriate response.

Afterward, I went into Lott's office and shut the door. He was my executive director, a man I'd known for years in multiple capacities. After a former provost hired me to work in the CITE, Lott and I developed systems to contextualize the different facets of our relationship. "Put on your teacher hat," I'd say. Or "your administrator hat," "your budget hat," "your *I am your boss* hat." Other hats he wore included: writer, editor, activist, faculty adviser to the student LGBTQIA organization, faculty adviser to the Office of Multicultural Affairs, community engagement liaison, instructional designer, the administrator who fought to enact college-wide name change policies for trans and gender nonconforming students, the administrator who fought for adjunct faculty to be paid for professional development, the administrator who fought to center student voices in strategic planning, and the person who reads more poetry than anyone else I know. Also, on a more personal note: he's family, my son's uncle, my village. I hope everyone is so lucky.

"Put on your *I've been in higher ed administration my whole career and I know how to dance the dance* hat," I told him.

"Cinema Arts and Sciences?" he said.

"IT'S SINGULAR," I yelled.

He sat behind his desk, looking very calm as I paced around his office and went off about wasted time and resources.

"How do you keep it together?" I demanded. "I want to stand on a chair and scream."

"I hear you," he said.

"I agree with you," he said.

"Will any of that help students?" he said.

I sat.

For the next hour, he taught me to control my face. We talked about meditation. About counting: down from a hundred, or the number of letter *e*'s in the document in front of you. "Find what works for you," he said, and after some trial and error, I landed on music, playing songs in my head whenever things got hard. When our instructional technologist, a man with a doctorate in online education, was told he had to teach faculty how to turn on their computers, I played "We Built This City" by Jefferson Starship. When our director, a brilliant woman from the writing center at Princeton, had her office taken away for the audacity of going on maternity leave, I played "Barracuda" by Heart. When we were told to make videos for the provost the day after being forced to fire our digital media specialist, I played "I Want Candy" by Bow Wow Wow.

Here's the truth: I needed the job. My husband had just left his to blog full-time. We were trying to build back our credit after near foreclosure. It was pre–Affordable Care Act and my son and I had preexisting conditions.

This is true, too: I loved the school, the same one that had changed my life years earlier as a student and where I'd taught on and off for a decade.

When do you fight and when do you cope?

For me, music helped.

Right up until it didn't.

*

"Think of it like this," I'd say in CV workshops. "If you're an actor, you show up at auditions with a one-sheet resume stapled to the back of a head shot. That's industry standard. A CV is standard in higher ed; it's how we show what we've done and what we're doing. No single template will work for everyone, however, there are conventions." I'd write examples of expected categories on the whiteboard—name, contact information, education, professional history—and immediately, hands would go up. At first I was surprised at some of the questions. Now I look at them through the lens of my own privilege. I held a seemingly secure position and was paid a living wage. I had health insurance that covered my family. I knew my work was valued and was beginning to understand how rare that was.

CATEGORY: NAME

I read that search committees are biased against women, against people of color. Should I use my first initial instead of my full name, like S instead of Sarah? Should I use a whiter version of my name? Joe instead of Jose, Rose instead of Rosanna?

CATEGORY: ADDRESS/CONTACT INFORMATION

Can this part be left out? What if people on the hiring committee are biased against my neighborhood? What if they don't want to hire within the city? What if they're only hiring locally? How will my paperwork get past HR if people stop reading at the second line?

CATEGORY: EDUCATION

I don't have a PhD. I have a PhD but it's in the wrong field. I have a PhD in a different field, but thirty years of professional experience in what they're looking for. Should I go back to school? I don't have a masters. Can I move the education category to the back of the document? I don't have a BA but I have twelve Grammys and a Whiting and a Nobel Peace Prize and 7 million followers on Twitter. Does any of that matter?

EXPECTED CATEGORY: PROFESSIONAL HISTORY

Can't I just say "faculty"? Students don't know the difference. What do these titles even mean?

*

Over the course of my career, I've had many titles: adjunct faculty, visiting faculty, visiting lecturer, visiting writer, contributing writer, writer, artist, teaching artist, teacher, instructor, administrator, assistant director, founding director, director. My title at the CITE was associate director of Faculty Development, which I adored because of the mail I'd receive addressed to Megan Stielstra, Ass Director. I'd circle the ASS in red marker and hang the envelopes over my desk. Lott thought this was hilarious. We'd drink coffee and look at the Ass Wall and laugh. Jesus, we needed to laugh. His boss, I wager, would not find it humorous. She was the associate senior vice president/assistant provost of Academic Student Engagement Instruction and Learning and Global and Recruitment Initiative Affairs, which is a position held by several people at this particular college and maybe other colleges, too. I'm not sure. I'm not sure what she did. I am sure that she didn't know what I did. She'd never seen my work, though it was her job to defend that work. Honestly? I don't think she knew who I was. She called me Maggie a lot. There was a Maggie in my office, as well as a Megan. We had completely different jobs, completely different personalities, and looked nothing alike; still, it was fascinating how often we were confused for one another and subsequently blamed for that confusion, as though us both having *m* names was a personal affront.

Real talk: Maggie is totally rad. Being mistaken for her was an occasional annoyance, at worst. I would take that annoyance and multiply it times ten, times a hundred, a thousand and try to imagine what faculty, staff, and students from non-privileged identity groups—people of color, queer and non-binary folks—experience in college on the daily. How often they're expected to control their faces both in and out of the classroom. I am in awe of their patience. I am furious they have to exhibit such patience. The CITE tried to run workshops on unconscious bias, on gender and racial justice. We talked with faculty about the changes they wanted to see and how we could support those changes. We talked with students, particularly a group called the One Tribe Scholars who received scholarships through Multicultural Affairs to work on issues of social justice on campus.

The hardest part: how to get the people who needed to be in the room into the room.

In the end, we were told to run more CV workshops.

*

If you have a position fortunate enough to include your own office, be grateful for the door. With a door, you can cry and rage in peace. My cubicle was visible to fifty-some people I barely knew and of course it's not—what's the word—appropriate to exhibit frustration in a public workspace, especially for a woman, especially in the academy where emotional response is discouraged and emotional labor is discounted. Be professional. Be removed. "Don't care so much," I was told again and again, and I'd bite my lip and play Fiona Apple screaming in my head.

I'm not sure when I started riding the elevator: up to the fourteenth floor, down to the first. I was alone. I'd sing. I wanted it out of my body. Liz Phair, Joan Jett, Pat Benatar. My mic was my fist. I wasn't very good but who gives a shit. Who gives a shit if somebody hears. Who gives a shit about the annual report, the programs cut, resources cut, services cut, departments merged, teachers losing classes, teachers not paid, tuition rising, students dropping out, faculty blindsided, everyone asking what the hell was going on.

"I'm sorry," I said, again and again. "I don't know."

Up to the fourteenth floor, down to the first.

*

Our health insurance switched during budgetary restructuring and all staff were required to attend a four-hour-long PowerPoint presentation, the kind where you're given a handout, and that same handout is projected onscreen, and someone reads the projection aloud word by tedious word. The font was so small. It might have been comic sans. I wanted to climb out of my skin. Instead, I got online and asked Twitter if I could start a dance party.

I have speakers in my backpack, I typed, counting characters to 140.

I could play Beyoncé. I could move the tables out of the way. I could throw the PowerPoint projector//

out the window. Maybe, if I really put my shoulder into it, I could throw it into Lake Michigan.//

That's why we have a lake, right?//
Somewhere to throw PowerPoint projectors?

Eventually, Microsoft tweeted back at me. There were lots of exciting ways to use PowerPoint! They were here to help! Would I like suggestions?

Afterward, I had pho with my dear friend Bobby, an adjunct teacher I'd worked with for years. I spent the entire hour bitching. Four hours for fucking health insurance. Bobby listened as I went on until, midsentence, I remembered that he didn't have health insurance.

The shame was a hand grenade.

I'm no better, I thought.

I have to be better.

I'll be better.

*

Staff members who'd been teaching part-time classes for years, some decades, were told they were no longer qualified.

I'm not a good enough writer yet to explain what that did to my heart.

Lott put on his executive director hat and walked into the then provost's office. He told her that if his staff wasn't allowed in the classroom, he would resign effective immediately. "How can we stand in front of teachers and talk about teaching if we're not teaching ourselves?" After that, CITE members were included on a special list, but as enrollment continued to drop and adjunct faculty continued to lose classes, I asked to be removed.

Me standing in front of students meant a friend might not work. A friend might not eat.

But I missed working with young writers.

I started teaching night classes at a different college.

*

In line for coffee I chatted with another senior official and she asked if I had a minute to talk. Couch space was limited, so we sat on a bench in the lobby of a nearby hotel. I don't remember the buildup to this particular discussion, the X and the Y before we got to the Z, but suddenly she was explaining the importance of being an "advocate" as opposed to an "activist"—to improve the system, not fight it. "Students look up to you," she said.

In my head I played the Sinéad O'Connor song that starts really quiet and gets more and more furious until she yells: "I'd kill a dragon for youuuuuuu." I explained, very calmly, how every week I work with teachers there voluntarily to learn about inclusivity. How every week I try to get teachers in the room who don't think they need to be there and until the administration makes it mandatory and—for part-time faculty—paid, it's never going to happen. How every week I meet with a former student whose current professor regularly misgenders him. "We talk strategy," I tell her. How to engage this professor. How to "advocate," but still, it keeps happening, and now this student is dropping the class because truly, is this his job?

No, it most certainly is not.

"We have systems for that," said the senior official. "That teacher should be disciplined."

At the beginning of every workshop in the CITE, we ask faculty to introduce themselves with their names, departments, and gender pronouns. I thought of all the teachers who'd thanked us for bringing the needs of trans and gender nonconforming students to their attention.

"Shouldn't that teacher be taught?" I asked.

She had to go then. It was a really busy day.

*

We received an e-mail from an interim somebody or another stating that after robust dialogue in accordance with nationwide best practices, adjunct faculty would no longer be paid to conduct student conferences.

Within the hour my inbox was full of fury.

There is a difference between common practices and best practices.

*

When they told me the One Tribe Scholars were losing their scholarships, I rode the elevator for fifteen minutes yelling the greatest hits of 1985. I was doing "A Hazy Shade of Winter," full-on air guitar and "hang on to your hopes my friend" when the door opened to a senior college official. I don't remember which. There are so many of them. They have so much power.

I put down the guitar.

He got on.

The doors closed, the floors counted down.

"Simon and Garfunkel?" he said.

We didn't look at each other.

"The Bangles," I said.

We stared at the numbers over the door.

"Oh," he said, and that was it.

The doors closed, the floors counted up, and I choked on everything I should have said, but didn't, and tried to do, but couldn't.

*

"Here's what you do," I'd say in CV workshops. "Put on a movie, one you've seen a thousand times and don't need to pay attention to. I use *Alien 4*, but you do you." Everyone laughed. Jesus. We needed to laugh. "Then grab a notebook and start making lists: jobs you've held, classes you've taught, presentations, publications. You'll start noticing patterns—categories—specific to your own unique experience. Some examples: exhibitions, performances, honors, service—"

Hands were already in the air, asking about possible categories, and we talked about the things that make up our lives as educators that don't fit on this course of one's life.

Other Jobs: Jobs I Had to Support Myself While Teaching

Activities We Came Up with to Help Our Students Learn

Brilliant Teachers I Was Lucky Enough to Work With

Brilliant Students I Was Lucky Enough to Work With

Student Successes under My Guidance

Students I've Supported through Personal Difficulties

Personal Difficulties I Survived While Teaching

I looked around the room. I'd worked with many of these people for years. I'd observed their classes. We'd sat in coffee shops for one-on-one consultations. I'd been to their art shows, met their families, watched them step outside their comfort zones to help their students learn. I knew who was fighting cancer, who had lost children, lost homes, who was getting businesses off the ground, winning awards, on deadline and still showing up for students, ten-, twelve-, fourteen-hour days with limited resources and little respect.

Tick tock my ass.

That night, I opened the document with own my CV, scanning down to the lists of faculty development workshops I'd designed over the past seven years.

I thought about how long it had been since I woke up to a job I loved.

I thought about the course of my life.

In the weeks that followed, I applied to some teaching gigs at other colleges. I sent out a few pitches. I outlined a book proposal, a collection of essays about fear.

*

Lott was told by some senior vice somebody that he had to lay another one of us off, the CITE's third team member in five years. When he protested, he was told to stop caring so much, that it was a detriment to his leadership capabilities.

He did not do a very good job at controlling his face.

"Put on your *what the fucking fuck* hat," he said, slamming around his office. I looked at the art hung on his walls, work that students had made for him over the years. I thought about my coworkers, how brilliant they were, how undervalued. I thought about the job interviews I had set up, and the Affordable Care Act, and the ad revenue from my husband's blog, and how much I missed being in the classroom.

Shhhhhhhh, can you hear it?

If I had the ability to embed video into this essay, it would be the scene from *The Hunger Games* where Katniss Everdeen throws herself at the stage and yells, "I VOLUNTEER AS TRIBUTE!"

Afterward, I took Lott with me to the elevator.

Up to the fourteenth floor, down to the first.

We sang "Desperado," by the Eagles.

*

The last faculty development workshop I facilitated was on the book *Teaching to Transgress* by bell hooks. Afterward, people stayed in their chairs. *Could we keep this conversation going?* they asked. Could we pick it back up next semester assuming they have classes next semester, no one was telling them, and how can you live like that, not knowing if you'll have a job next month and—

I lost control of my face.

"Do any of you watch *The Vampire Diaries*?" I asked. "Forget it, it doesn't matter. There's this character named Bonnie. She's a witch, a portal to the underworld. When people die, they have to pass through her body and she can feel them as they move through her, the fear and anger and confusion from hundreds, thousands of people." Everyone looked at me, baffled, but I kept going. "She carries all of it with her and she wants to make it better but she doesn't have that kind of power and I just—I tried. I really tried."

With that, I stood up, went to the bathroom, and sobbed.

Jesus, it felt good.

It felt, finally—true.

*

I took the L downtown for my exit interview, lugging a hiking backpack to pack up my desk. I got off at the Red Line and climbed the stairs to the sidewalk, then a block to Wabash and another block to the corner of Michigan and Harrison. I'd walked that walk ever since I arrived as a transfer student in 1995. When you say Chicago, this is what I'd see: the grid quick frying in the July heat; the white carpet of the park in December; the underside of the Brown Line pounding overhead; the impossible beauty of the lake in the morning, blue or green or gray; taxis blowing speed limits and running red lights up and down Michigan Avenue; and commuter students smoking on the sidewalk. We had tattoos and wild hair, all ages and races and genders, we liked girls and boys and neither and both, we made our own clothes and worked six jobs and studied our asses off and wrote stories and poems and music and films. And yeah, sure, we drank too much, and sure, fine, there were drugs, and of course, *of course*, sex, either way too much—thighs chafed and aching up the sidewalk—or never enough. *Why doesn't he liiiike me why doesn't she waaaant me why don't they looooove me*. And sometimes it was perfect, sometimes sad, other times dangerous in ways

I couldn't articulate for years, and sometimes—most of the time, all of the time—awkward, but we were living, trying on identities, trying on relationships. We put them on and took them off like sweaters. I lived in the Ukrainian Village, then Wicker Park, then Logan Square pre–farmer's market, then Humboldt near the park and Uptown. I worked in Little Italy and River North carrying cocktails and cheese plates and homemade pasta, then the Bongo Room pouring mimosas, teaching in Hyde Park, Ravenswood, Cabrini-Green, the Loop—I can't remember everywhere anymore. What I do know is this: in all the moving, all the rushing, twenty years trying to get by and make it and make things in this beautiful, complicated city, the only place that remained constant was that corner at Michigan and Harrison.

So much of who we are is tangled in place; a country, a city, a corner.

I didn't want to be there anymore. That doesn't mean it was easy to leave.

It was the day before the holiday weekend and campus was pretty much empty, just me and a woman in Human Resources there to "separate me from the college." That's the vernacular: separate. I think it's appropriate, literary even. I'd been there a month shy of twenty years.

We sat across from each other and she went over her checklist: insurance, vacation days, 401K. She was good: kind, informative, and apologetic without being gross. I wondered how many of these separations she had to facilitate. Hundreds? Thousands? She should get a raise.

So should your teachers.

At the least, a living wage.

At the very least, to know they were valued.

*

It didn't take long to clear out my cubicle. I stuck books into my backpack and then opened the file cabinet. The top drawer was stacked with CVs I'd worked on with faculty. I thumbed through.

God, so much talent.

So much accomplishment, so much experience and expertise.

You see that, right? When you look down the course of your life? What you've made, how you've helped, how you've loved?

It had been a while since I opened the bottom drawer—the stories and essays and poems I used in my own writing, my own teaching. Gabriel García Márquez. Toni Morrison. Dorothy Allison. Kafka. Louise Erdrich. Joan Didion. Annie Dillard. Joy Harjo. I sat on the floor and read for a bit, trying to remember who I was before this all started.

I liked that girl.

You always knew what she was thinking. You could see it on her face.

I stuffed everything from the bottom drawer into tote bags and left the top drawer as it was. Then I stood up and looked at the cubicles around me, four rows of six, an empty sea. I was the only one there. So I played some music in my head and did a little synchronized swimming, leaping through the aisles, fanning my arms like peacock feathers, sinking below the desk line and springing back up.

The Blogger's Wife

1) I have an idea.
2) It's called "The Blogger's Wife."
3) I'm not sure if it's a story or an essay.
4) It's about a woman who's married to a blogger and if someone leaves a shitty comment on one of his posts she tracks down their IP address and shows up at their house and duct-tapes them to a chair and gives an inspirational yet scathing monologue about what it means to be a decent human being.
5) Naturally, there's an orchestra playing in the background.
6) It's probably a story.
7) I haven't worked out the dialogue yet.
8) I haven't worked out the duct tape thing, either.
9) What if she drugs him? Dares him? Hypnotizes him? Challenges him to an arm wrestling competition and if she wins, he tapes himself to the chair and if he wins—forget it, he can't win. I'll make her super strong. She's an Ultimate Fighter. She's Ronda Rousey. She has a 170 IQ like Judit

Polgár and dazzles him with logic. She shows him a PowerPoint presentation with visual data about the long-term effects of cyberbullying and/or bullet points on how to remain civil during Internet discourse. She plays the episode from *This American Life* where Lindy West interviews one of her online harassers. She reads aloud from "The End of Empathy" by Stephanie Wittels Wachs: "We're never going to get anywhere if we continue to treat each other like garbage." She reads poetry, like this from Mary Jo Bang: "We pretend we forgot that we said we'd be kind."

10) Whatever she does, for sure it won't be a gun.
11) Enough with the guns.
12) A year or so ago, my son told me he didn't want to see any more kids' movies where the parents—be they human or animated Pachycephalosaurus—died in the first ten minutes. *Finding Nemo. The Lion King. How to Train Your Dragon. Tarzan.* He wasn't traumatized. He was bored. "Don't they have other ideas?" he asked.
13) That's how I feel about guns.
14) (Maybe I'm a little traumatized.)
15) (Maybe we're all traumatized.)
16) I want to imagine other solutions.
17) I want to imagine other possibilities.
18) I believe our capacity for imagination is stronger than our capacity for fear.
19) Fire, spoken language, the domestication of plants, the wheel, the alphabet, the pill, paper, electricity, anesthesia, engines, telegraphs, telephones, democ-

racy, penicillin, the airplane, DNA, the Internet, refrigeration, period-proof underwear, rocketry, self-driving cars, man on the fucking moon.

20) We can do better. We can.
21) When I'm feeling optimistic, I think of the common-sense stuff: universal background checks, tighter enforcement of existing laws, smart gun technology to reduce accidental shootings by children, outright ban on assault weapons, demilitarize the police, greater investment in mental health care and education and our poverty-stricken communities.
22) When I'm feeling cynical, I think we've proven time and again that we're not responsible enough to handle guns and we should repeal the Second Amendment in its entirety.
23) Regardless, we need more discussion about anger management.
24) Domestic violence.
25) Toxic masculinity and how we raise our sons.
26) I think this is an essay.
27) I'm afraid to write this essay.
28) I'm afraid that I won't say it right.
29) I'm afraid that what I say will make weirdos threaten me on the Internet.
30) Sometimes weirdos on the Internet say they want to rap you and it takes a second to realize they forgot the *e*.
31) Sometimes weirdos on the Internet write very long letters that they send to the e-mail addresses on your personal website and the college where you

teach and the college where you used to teach and the theater company where you work and your various social media accounts including Google+, which you hadn't checked in years and you're like: *Holy shit, dude, how do you have time to go through all of those directories?* I didn't have five minutes this morning to finish a bagel.

32) Sometimes weirdos on the Internet show up at the UPS store around the corner that your husband listed as the mailing address for his art blog back when he ran it out of your apartment. They sit there, on the curb, waiting. They want to show him their work. They want him to write about their work. They want him to sell their work, and you wonder, what if they knew where you lived?
33) *What if* is a dangerous game.
34) To be clear: the vitriol I receive is minor compared to that of other women writers, mostly women of color and queer and trans women whose work challenges me and moves me.
35) Still: bullshit is bullshit.
36) Here's a story:
37) A few years ago, my building caught on fire, those precious few moments in the chaos to get your family out alive.
38) I wrote about it for the *New York Times*, my first publication on so wide a platform.
39) After I filed the rewrite, my editor asked if I had any thoughts about Chicago's upcoming mayoral election.
40) My first reaction?

41) Fear.
42) Not excitement that he wanted my work.
43) Not gratitude for the opportunity to serve my city.
44) Not relief at the paycheck because hi, we needed it.
45) Fear.
46) I tried to talk myself out of it.
47) *I write stories, not essays.*
48) *I write personal essays, not political commentary.*
49) *I write political commentary to perform, not publish.*
50) While I engaged in this self-sabotage, my fire essay went live, both print and online. Within hours, there were hundreds of comments and messages and e-mails.
51) How I was stupid.
52) How I was fat.
53) How I'd put my child in danger.
54) My favorite: THAT'S THE WRONG WAY TO SAVE YOUR LIFE!!!
55) Three exclamation points.
56) There were supportive responses, too.
57) Thank you for those.
58) Note to self: lift up the good stuff.
59) I called my mother and made her promise not to read the comments.
60) (Idea: "The Blogger's Mother-in-Law." It's about the mother of a woman who's married to a blogger and if someone leaves a shitty comment on one of his posts she somehow finds *their* mother and sets up a parent-teacher conference and the three of them talk about how to say and do kind things.)

61) There is a liquor store across the street from my apartment. I bought a bottle of Maker's Mark. I bought fancy bitters and sugar cubes and an orange. I called my friend Amanda and she talked me through making old-fashioneds the way that she makes old-fashioneds.
62) She makes the best old-fashioneds.
63) I drank one very fast and a second one very slow.
64) Then I had—
65) Let's call it an epiphany.
66) It doesn't matter if the work is personal or political.
67) It doesn't matter if it's a story or an essay.
68) Some people will come after us no matter what we say.
69) We might as well say things that matter.
70) Audre Lorde: "We can learn to work and speak when we are afraid."
71) I wrote my editor at the *Times* and said, *Yes, I do have thoughts about my city's mayoral election*. The next day I chose a thousand careful words to say that Rahm Emanuel is dangerous and why are we voting for him, Chicago?
72) We still voted for him, Chicago.
73) We can do better, Chicago.
74) We can imagine the city we want to live in.
75) We can imagine this whole goddamn world.
76) Recently I was on a panel at a writing conference about the essay in the age of the Internet. In the Q&A that followed, a woman asked me and the other two female-identified panelists if we feared

for our lives. She mentioned a local public radio host, also female, who took a leave of absence because of ongoing threats of violence.

77) Another conference, another panel, this time about writing essays in the hopes of changing the world. A woman from the audience came up to me afterward and asked how I protect my child. She has children, too, she told me. She was afraid that publishing her work would put them at risk.

78) I am asked these questions all of the time.

79) Like it's normal.

80) *How do you take your coffee?*

81) *White wine or red?*

82) *Do you fear for your life?*

83) If you find that surprising, I invite you to sit very quiet and still and ask yourself why that is.

84) Every day I see women pushing back, in stories and movies and real life, online and off, on the news, in the streets, boardrooms, and pages against words, fists and Photoshop, in public and private forums from schoolgirls to legislators to performers. I'm not talking about criticism or disagreement or intellectual debate, all of which I think is good. I'm talking harassment: *bitch* and *cunt* and threats of violence, threats against our children, our bodies, our livelihoods.

85) Yes, I am afraid.

86) The truth of that makes me want to set the walls on fire.

87) Let's set the walls on fire.
88) In 2015, the music critic Jessica Hopper asked women and people from other marginalized communities to share their "first brush with the idea that they didn't count." There were thousands of replies, throwing a bright light on sexism in music including people who'd left the industry altogether for their own physical or emotional safety. "What songs or albums could we hear if people weren't being told they aren't supposed to be here?" Hopper told *the Guardian*.
89) What songs or albums—
90) What technological innovations or scientific discoveries—
91) What policy advancements or philanthropic endeavors—
92) What artistic or athletic achievements—
93) What stories or essays—
94) Is this a story or an essay?
95) It doesn't matter. Here's what I want it to say:
96) You are supposed to be here.
97) You are needed.
98) We need you.
99) We're imagining the world.
100) What if you're the one who saves us? The one who finds the cure? Deactivates the bomb or gets us to Mars or unites us as one? You're an Ultimate Fighter. You're Ronda Rousey and Judit Polgár and your genius is off the charts. You design the tech to protect us from bullshit and the drugs for pediatric cancer and a moon-size magnet

that sucks metal into space. You play Symphony no. 9 in D Minor and we remember the beauty in the world. You read aloud from "The Girl in the Cabinet" by Melissa Chadburn: "There is a child somewhere — a girl — and maybe she will pick up a book or peruse the internet and she will find your words. And in your words she will discover a world of the possible and she will climb out of the cabinet and she will put down the razor." You read us poetry, like this from Joy Harjo: "But come here, fear/ I am alive and you are so afraid/ of dying."

What Belongs to Us

When do you think about your own privilege?" Dia asks over the phone.

I wish we could have this conversation sitting across from each other. I wish we could have every conversation sitting across from each other, but she lives in Oakland and I'm in Chicago. Your best friend on the other side of the country is your body without any breath.

We talk every week, usually stuck in traffic; me headed south on Lake Shore Drive, her headed east on the Bay Bridge. We talk about our sons, our partners, our mothers. Should I get bangs—*yes.* Buy those shoes—*yes.* Stay in higher ed—*hmm.* We talk about Chance the Rapper and *Broad City.* We talk about that dumb thing on Facebook and that awesome thing on Pinterest. We talk about restorative justice and radical pedagogy and our kids' elementary schools and the respective cysts on our respective ovaries. She is the person I cry with; sometimes, when I hear her voice, I start sobbing, which is both inexplicable and exactly right. When we lived together in our twenties, she made me do yoga every morning. She got me out of the house, away from the computer. She taught me the word intersectional and gave me books on feminism and racial justice.

This is not a one-time conversation.

We've been having it for years.

*

I tell her I'm writing an essay for white people about privilege and responsibility. "Can I invite black readers to sit this one out?" I ask. "Like, put their feet up? Relax?"

"Oh my god, please," she says. "Make sure you tell us to pour ourselves some wine."

*

"When do you think about your own privilege?" she asks. Granted, the question can run in many directions—my privilege in being cisgender, heterosexual, middle class, American, able-bodied ("temporarily abled," says my friend Maria), with access to education, housing, and job opportunities and the multiple intersections therein—but I don't need her to specify. I know what we're talking about. We're talking about whiteness.

Dia is the education director at a nonprofit that helps people learn about unconscious bias and racial inequity. She facilitates workshops around the country, mostly for white audiences, teaching the history many of us never learned in school and, for the most part, haven't tried to seek out on our own. I've sat in her classrooms, fifty, a hundred, five hundred people, listening as she outlines the accumulated advantages and disadvantages that contribute to systemic racism in the United States. Some examples: The

1830 Indian Removal Act. The Naturalization Act of 1790. The Social Security Act of 1935. The National Labor Relations Act. The 1956 Federal-Aid Highway Act. Jim Crow. She connects the dots between past and present—the current systems of white supremacy that many of us, like me, benefit from every day. The courtroom. The classroom. The boardroom. In the media. Housing. Health care. Transportation. Criminal justice.

It's a lot of information. It gets uncomfortable: blame, shame, fear.

Dia is also a yoga teacher; she shows people how to breathe.

*

If this information is new to you, don't worry. It was to me, too.

This is why Jesus gave us Google.

*

"When do you think about your own privilege?" she asks, and I tell her about a class I took in college with the storyteller Emily Hooper Lansana. She took us century by century through the history of the oral tradition: video, audio, and live performance. She got us out of the classroom and into the city. She made us stand up and try, and she built a space where we felt safe to crash-land. She was so smart. She was so cool. At first, I wanted to impress her; but soon I forgot about that and tried to make good work. I loved telling stories. I loved listening to stories. I saw how they could change the world.

That's part of being a good teacher, I think: inspiring people to work their asses off not for you or the class or the grade, but to be better.

Near the end of the semester, Emily showed us a video of Anna Deavere Smith performing *Twilight: Los Angeles, 1992*, embodying different people involved in the Los Angeles riots in monologues based on oral interviews. I remembered reading about Rodney King in the news in high school, but this was different. This wasn't statistics or politics or op eds. This was real people, real stories, a world I'd never seen. "Why didn't I know about this?" I said, incredulous, not yet understanding that it was my responsibility to look.

"Come with me," Emily said. We walked from our building on Wabash, around the corner, a block down Balbo, around another corner to Michigan Avenue, and into the school library. I'd been there a hundred times—assignments from teachers and research for advisers and books for class—but this felt different. It felt like when I was a kid: the library as a place of discovery, to follow my own curiosity, to crack open the world.

Emily gave me back the library.

The student work aide that day was taking a class on critical race theory. When I asked her for help, she literally rolled up her sleeves and said, "Let's do this."

W. E. B. Du Bois. *Assata: An Autobiography.* Angela Davis. bell hooks. Octavia Butler. Langston Hughes. Zora Neale Hurston. *Sister Outsider. This Bridge Called My Back. Black Boy.*

I took the books home, sat on the couch, and did what Emily taught me.

I listened.

*

At some point, our education no longer belongs to our teachers.

It belongs to us.

*

"When do you think about your privilege?" she asks, and I tell her about sitting in the audience at a literary conference, watching yet another panel discussion with all male participants and feeling that frustrating tangle of fury and boredom. Then, the very next day, I sat *on* a panel with all white participants.

My hypocrisy was a lightning bolt.

I interrogated my memory: readings I'd been a part of with all-white lineups; panels, anthologies, commissions; theater companies I'd supported, publishers I'd supported, magazines I subscribed to. I looked at my bookshelf. I looked at the lists of assigned readings on my syllabi. I looked at who I follow on social media, where I get my news, where I give my money.

It's easy to see the truth when we do the work of looking.

A friend gave me a template, which I now cut and paste from a file on my desktop:

> "Hey, thanks for thinking of me! I'm committed to events that prioritize diversity as an active practice as opposed to a talking point and would love to ask about the other collaborators insofar as race, gender, sexual orientation, background, etc."

I wonder how things would be different if everyone who's part of a privileged identity group asked that same question when they're invited to perform or publish or present, to join a company or faculty or board.

I wonder how things would be different if we paid attention to what was truly in our power. Example: A white man is convicted of rape and the media shares his senior photo and college swimming scores, as opposed to his mug shot and prior substance abuse. A thirty-two-year-old white Olympian pisses on walls and lies to international authorities and people say "boys will be boys," while a black child gunned down by the police is "menacing . . . a 12-year-old in an adult body."

I teach writers. It's on me to show them the weight of words, how they can perpetuate or elevate.

Privilege isn't blame or shame or fear.

It's responsibility.

*

Let's try an experiment: jump on the computer, log into Google and type in *racism in* ________________, and then include your job or aspects of your life. For example, I'd write *racism in literature* or *racism in higher education*. For my husband: *racism in visual art* or *racism in technology*. Our son: *racism in karate* or *racism in Pokémon* or *how to talk to your seven-year-old about racism*.

What about your city?

Racism in Chicago.

What about the systems you use every day?

Racism in housing, transportation, education, food.

What about the art and media you consume?

Racism in Hollywood, in reporting the news, in the music industry.

What do you notice?

And now—what can you do?

*

"When do you think about your privilege?" she asks.

We were at your house for Thanksgiving. The boys wanted to play in the front yard with plastic swords and squirt guns. My son didn't understand why your son wasn't allowed to be outside with a toy gun. They looked at you. You looked at me. You and I had a conversation that didn't involve speaking, and my son and I went for a walk. I told him about Tamir Rice. About Tyre King. I grabbed the air for words to explain, knowing that my heartbreak is a puddle compared to the ocean you swim every day.

*

I'm not saying anything new. Black women have been asking this of us for generations.

*

"When do you think about your privilege?" Dia asks. Traffic is still crawling, south on Lake Shore, east on the Bay Bridge. We talk about our sons. Our partners. Our mothers. Should I go to Jamaica—*yes*. Write that essay—*yes*. Cover the gray—*who cares*. We talk about *13th* and Eula Biss. We talk about the awesome thing on *Rent the Runway* and whether or not to engage on Twitter. We talk about segregation and decolonizing yoga and our kids' elementary schools and the new curriculum she's writing for parents and caregivers of white children interested in age-appropriate discussions of racial inequity. She wants to know how I feel as a mother, as an academic. She searches my stories for questions, ones to bring back to her white audiences. She's making our country better for her kid and my kid and everybody's kids, and she wants us to join her in that work.

It's not a one-time conversation.

I'll be having it for the rest of my life.

forty, or Optimist

31

I told the Realtor that I was scared of Realtors—of buying property, owning property, credit checks, debt of any kind, furnaces that explode a week after the warranty expires, buyers' markets and sellers' markets and markets in general and niche vocabulary like *equity* and *escrow* and signing lots of documents that I don't understand but am bound to, like cell phone contracts or gym contracts, and if I'm nervous about contracts that involve, like, a hundred bucks, then why would I dick around with not only my entire life savings but also tons of money that I don't even have? "It's like when my Isuzu died," I told him. "I went to the dealership and I was like, look: I only have X amount of money so don't tell me later that I have to pay a zillion dollars in state tax and warranty and what all, just give me a car that has decent mileage and won't fall apart when all these crazy Chicago drivers bang into it parallel parking, you know what I mean?"

He did.

He specialized in historic buildings: vintage architecture, nontraditional layouts, the opposite of pop-up McCondos.

Also he was patient, explaining everything fifteen times in layman's terms, all of which I wrote down in specific detail although later, when I read back over those notes, I didn't know what the hell they meant, which is very similar to reading over stuff I wrote back in college when I was drunk.*,† Also, he jumped up and down on hardwood floors to make sure there wasn't excess moisture, which is apparently very bad. Also, one time, when he was driving us around from condo to condo, some guy cut him off in traffic and my Realtor pulled up next to him at a red light, rolled down his window, and said, "YOU MUST DRIVE BETTER." He used the same tone of voice anyone else would've used to cuss somebody out. Also,

---

* I do not advocate writing drunk. If you must, please clean it up before you turn it into your editor and/or professor.

† I do not advocate writing drunk. One time I was at a bachelorette party and I wore very tight jeans with one back pocket full of cash for strippers—Zorro was my favorite—and in the other pocket I kept the notes I took to write about Zorro later. When I woke up the morning after, I first had to wait until my eyeballs stopped vibrating. Then, coffee. Then, I wandered around till I found my pants: one pocket full of bills, the other, empty, which meant that I'd put notes instead of dollars into Zorro's G-string. I have since mythologized those notes to Pulitzer-winning heights and deeply regret their loss.

As I see it, this anecdote holds the following morals:

1) I owe Zorro. Stories don't pay rent. And hey, while we're on this subject:

A) Pay service workers. Gratitude doesn't pay rent.

B) Pay interns. Experience doesn't pay rent.

C) Pay writers. Exposure doesn't pay rent.

D) Pay everybody on time. Rent is due right now.

2) Take notes in a dedicated place, not on the backs of individual order slips that you stash somewhere for later and subsequently lose forever. We need those notes. We need those memories. We need those versions of ourselves. Per Didion: "We forget all too soon the things we thought we could never forget. We forget the loves and the betrayals alike, forget what we whispered and what we screamed, forget who we were."

he plays in two bands and is super cool to have a beer with. Also, he found our new home, which, at the time, I deeply loved.

31

New neighborhood, new condo: in Uptown across the street from the Aragon Ballroom. Our street was full of tour buses: Pixies and Yeah Yeah Yeahs and Bob Dylan and the Flaming Lips, guys in orange vests directing traffic as bands tried to parallel park, fans lined up down the street for the Aragon and the Riviera and the Green Mill underneath the Red Line stop at Lawrence. Due north: the West Argyle Street Historic District, or Little Saigon. Due east: five blocks of Lawrence Avenue ending in the Montrose Dog Beach, where we went religiously every Sunday so Mojo could run it out. Due south: an empty parking lot under an enormous billboard that changed ads every few months, mostly for upcoming films (*Little Miss Sunshine, Casino Royale*) and various cold medications (we immediately got sick). And high above it all, eight hundred square feet of a top-floor walk-up with a balcony and a turret and a refrigerator that had a little button and when you pressed it, ice came out, *my* ice, ice that I owned, ice bought and paid for with the American dream, i.e. Monopoly money and your firstborn child. It felt exactly like that Lorrie Moore story "You're Ugly, Too": Zoë buys a house. She keeps going into the basement because "it amused her to own a basement. It also amused her to own a tree. The day she moved in, she had tacked to her tree a small paper sign that said Zoë's Tree."

I didn't own a basement.

I didn't own a tree.

Somehow, that made it less scary.

31

A year after going off the pill, alone in a single-occupancy restroom at the Uncommon Ground on Clark Street. I peed on the stick. I waited the three minutes.

31

Four hours later, I was on a table at the gynecologist's office. I'd made an appointment after three plus signs appeared in three tiny windows of three separate pregnancy tests. The soonest they could see me was after lunch, so I went to the table where Jeff and I met up to write, twitching from the effort of not telling him. My whole body shook. "No more coffee for you," he said. "Okay," I said, drinking water instead, glass after glass to keep my hands busy.

Recently, I went back through my journals to find what I wrote that morning. At the top of the page is the date, followed by a single word: *please.*

When I got to the doctor's, she had me pee in a cup and tested it for HGC. The results: not pregnant.

I'm nineteen years old, there on the table.

Thirty-one, on the table.

It doesn't matter how old you are, how well you know your body.

I showed her the three plastic sticks with the pink plus signs. "How about that?" she said, and did a blood test, which verified the pregnancy.

Apparently I'd drank so much water that there wasn't pee left in my pee.

32

I didn't know I was supposed to be terrified about bringing a child into this world until I went to the eco baby store. I was there to buy paint. Our friend Kat was making the baby a mural, two weeks on ladders and a twisty forest climbing his walls. At the time, we were desperately broke, watching the value of our home plummet with the recession and trying, like many of us raising children, to figure out how we'd manage both a mortgage and child care.

The saleswoman brought me the paint and asked if I had a crib mattress.

"I—"

She listed the ways mattresses would kill my baby and pointed at one that was safe. It cost over a thousand dollars. I went home and read about VOCs. About SIDS. About entrapment. About vaccinations and cord blood and cryogenically freezing your baby's teeth to preserve stem cells in case your baby gets cancer. Did you know formula will give your baby cancer and that hat will give your baby cancer and what do you mean you don't breast-feed, you're killing your baby and how can your baby live without a Bumbo and a Boppy? I read about life insurance and health insurance and auto insurance, renter's and casualty and disability and college savings and you have to pay for college or your child will be in debt forever and without college they won't get a job ever and they'll live with you forever and you can never retire and how will you retire and why are you eating that chicken it will kill you GMO HMO PPO and you're already thirty? You should freeze your eggs. You should wear this band thing to help you get your abs back. You have to get your body back or no one will love you. Here: makeup for new moms, meal replacement for new moms, moisturizer for new moms that can freeze your cellulite and also—in teeny tiny print—*maybe kill you*. And in huge bold letters—*but don't think about that! Think about no cellulite! No woman wants cellulite! Cellulite cellulite cellulite!*

I decided to take a break from the Internet.

32

Christopher set up a meeting with a financial adviser to discuss college savings options. I was mystified by such forward thinking. *I will eat a bagel sandwich. That* was all the planning I could handle.

So. Math! My kid will be of typical college age in 2025, which means, at the current rate of inflation, four years of undergrad will cost NINE HUNDRED BILLION DOLLARS.

For real though: we sat down with the financial adviser in his swanky financial adviser's office, and by "sat down" I mean wedged my pregnant ass into a teeny little chair with arms—

Dear People Who Have Offices and Therefore Probably Chairs: Get the ones without arms. Not everyone can fit in the ones with arms. Some of us are fat. Some of us are pregnant. Some of our bodies don't work the way yours does. Truly, this is not hard! There are so many options! I believe in you!

—and he explained with a totally straight face that in order to pay for college outright in 2025 we'd have to save, ballpark twelve hundred dollars a month for the next eighteen years. I burst out laughing. And my laugh—it's obnoxious. What's that line from *The Catcher in the Rye*? "If I ever sat behind myself in a movie or something, I'd probably lean over and tell myself to please shut up."

In that moment, laughter was the logical response, easier than facing an impossible reality, better than fearing an inevitable failure.

32

Christopher's office threw us a baby shower. He was a web developer at a financial firm in an enormous glass skyscraper in the West Loop. It was very fancy. The pay was good and there was downtime, enough to start an art blog on the side, but still, it was eating him alive. The company dealt with huge amounts of money, so much so that the government built a mirror office in a rural yet otherwise undisclosed location to continue operations in case the Chicago office ever blew up, which I think is the dictionary definition of cynicism.

There was a woman in the mirror office with Christopher's same job.

Her name was Christine.

Listen: I was grateful for that job with its shiny, pre-ACA health insurance that let us have our baby but man, was that place eerie. The baby shower was held in the conference room, long and narrow with a long, narrow table, the head of which was flush against the wall. The wall—get this—was a floor-to-ceiling video screen showing another long, narrow table in a long and narrow room in Texas. On that table, onscreen, was a cake that said: *Congratulations, Megan and Christopher!*

And on this table, in front of me, was another cake that said: *Congratulations, Megan and Christopher!*

The baby kicked the hell out of me the whole time.

My boy knows what's up.

32

At the infant care classes you take at the hospital, they give you a doll to cuddle and swaddle and practice changing diapers. I accidentally broke off its legs.

32

February 1, 2008, 10:00-ish p.m. Outside: snowstorm. Inside: chicken-fried steak from a bar around the corner called Fat Cat while watching *Lara Croft: Tomb Raider.* It was the bungee ballet scene, Angelina Jolie all ethereal in white gauzy pants dancing midair to Bach while shooting black-op assassins with machine guns, a soundtrack of bullets and techno and breaking glass. I remember getting up to pee, but when I peed, the pee didn't seem like pee.

I felt very dumb. One should know if one's pee is pee.

"My water maybe broke?" I said to Christopher.

He was playing online Scrabble, winning by a lot. "It's too soon," he said, which was true, we weren't due for weeks, but that morning I'd put both hands on my stomach and whispered, *I'm ready*. I imagined the baby and me communicating via a radio surveillance earpiece, the kind with the plastic telephone cord like Agent Smith wore in *The Matrix.*

"You're probably right," I said to Christopher.

Then I had a contraction.

I remember thinking that I'd never be a good enough writer to find words for the pain. I'd read about it on the Internet: "Really intense menstrual cramps," said one woman. "Overwhelming back pain," said another, and all I can figure is both those women must've been drugged out of their minds 'cause those words don't skim the surface. First of all, they're lowercase, and, believe you me, labor contractions ARE IN ALL CAPS AT ALL TIMES. Second, poor choice of adjective! Intense? Overwhelming? Try: WHAT THE GODDAMN FUCK. Try: ARE YOU FUCKING KIDDING ME? Try: TAKING A BULLET IN YOUR LOWER BACK A BULLET THAT IS ATTACHED TO METAL WIRES AND SOMEONE IS HOLDING THE OTHER END OF THOSE WIRES AND THEY RUN AROUND FROM YOUR BACK TO YOUR FRONT DRAGGING THE WIRE THROUGH YOUR INSIDES AND THEN YANKING IT OUT YOUR ABDOMEN AND IT'S IMPORTANT TO NOTE SINCE MY DAD IS A BIG GAME HUNTER AND HE'S PROBABLY READING THIS HI DAD! THAT THE BULLET IN QUESTION IS BUCKSHOT USED FOR LARGE GAME AND/OR MILITARY AND IS ACTUALLY LOTS OF LITTLE SHOTS INSIDE OF ONE BIG SHOT SO REALLY THERE'S A HUNDRED WIRES RIPPING YOU OPEN INSTEAD OF JUST ONE AND THAT HAPPENS EVERY FIVE MINUTES.

Poor Christopher. He'd read all the books, the ones that tell the new partner how to help the birthing mother through labor, and all of them talk about those first eight to twelve hours: holding her hand, walking her around the block, giving her water through a straw, timing contractions from hours apart to five minutes apart and then—and only then—calling the hospital, and here I was, five minutes apart from the get-go and, in the background, a soundtrack of submachine guns and death.

"Are you sure?" he asked.

I puked in the sink.

"Okay," he said, calling the hospital.

What happened next was forty minutes of Christopher digging out the car with a shovel, by which point I was covered in vomit and scared out of my mind. Then a slushy, snowstormy drive down Lake Shore, me turned sideways in the passenger seat with one foot on the windshield and my leg bent across the dash. I had several moments of profundity: No pleasure without pain. No joy without fear. Etc. The time between contractions was so mind-blowingly wonderfully breathtakingly glorious and I wouldn't have been able to experience such glory without the paralyzing pain coming every three minutes.

We arrived at the hospital, and I thanked the snowstorm for clearing traffic that night. I thanked the baby for waiting until after rush hour. I thanked Dr. John Bonica for inventing the epidural that would soon be mine. "I'm having a baby," I told the intake nurse at triage. "Right this very second."

She probably heard that line twenty times a day. "Sit down, dear," she said. "You have to fill out these forms and—"

I puked. Everywhere. Over everything. Puke on the floor, puke on the desk, puke on the pretty, newly renovated hospital wallpaper. It got on the documents I was supposed to fill out and the woman who called me dear. We looked at each other. Then we looked at the floor. Then—and maybe this is just me—you know how when you've had too much to drink, the only thing that could possibly make it better is to take off all your clothes and lie naked on the floor? I stripped in the waiting room. One boot, one sock, my winter coat, pants pulled over the other boot as I pushed through the door—I'd found a door! And down a hall—look, a hall!—then sweater, shirt, and underwear. When I finally got to the toilet I was wearing one Ugg. Christopher told me he followed the line of clothes.

They got me on a bed and a nurse came in. She was very nice, but, like the receptionist, thought the immediacy of this was all in my head. Then she took a peek, said, "Oh my goodness!" and disappeared. And within seconds I was on a gurney and down another hall. The delivery room was fancy, with all sorts of machines and a forty-two-inch flat-screen TV. *Beetlejuice* was playing. *Beetlejuice,* with Michael Keaton and pre-*Heathers,* pre-*Mermaids,* waaay pre–*Reality Bites* Winona Ryder. It was the dinner party scene where they all dance to "Day O" and Catherine O'Hara wears a single perfect, black leather elbow-length glove.

Dr. F. came in. FYI: if you're having a baby, this is your guy. Super cool, super calm. "Heeey," he said, like he'd just smoked a bowl. "Let's have this baby."

"Let's have an epidural," I said.

"Let's push," he said.

There was no way in hell I was doing it without drugs. No way, no how. It was just too much: the crying and the puking and all of it happening so fast, so immediate. "You already did all the hard stuff!" my friend Julie, my role-model mother, told me later. "Pushing is a relief in comparison to getting to nine centimeters!" But I wasn't thinking about any of that. I was thinking that I didn't want to have this experience hurting anymore. I wanted to have it joyfully. I wanted to . . . float.

"No, really," I said. Probably "said" is incorrect. Gasp? Yell? Plead? "I need an epidural."

"Try pushing once," F. said.

"That's fair," I said.

I didn't know that once meant three sets of ten.

Fuckin' doctors. Fuckin' personal trainers.

"Okay," I said after. During, I'd imagined that Ralph Steadman illustration where a guy's brain explodes upward though his skull and the ceiling drips with brain goo. "Epidural, please."

Finally, it wasn't about me anymore. It was about the baby. The baby. The baby. I hadn't thought about him in the past few hours—only myself—and I remembered that this little person I'd been talking to for months was here.

I didn't feel the episiotomy. I pushed five times. I watched Christopher's face. It was, for me, the most amazing way to bring a child into this world: calm, slow, knowing exactly what was happening because it was all over his face. He gripped my hand. He kept forgetting to lift my leg. When our son arrived, Christopher caught him and they yelled like wild men and I wasn't scared.

At least, not yet.

32

We are awake, the baby and me. The 2008 presidential primaries are on cable.

32

We are awake, the baby and me. *The Lost Boys* is on cable.

32

We are awake, the baby and me. *Saving Private Ryan* is on cable. I'd seen this movie a few times before and was always like: *No.* I was like: *You do not risk the lives of many to save just one.* I was like: *Are we saving Matt Damon again*? But this time, my kid in my arms, I was a sobbing mess. "YOU GET THAT BOY," I yelled at Tom Hanks. "GET HIM RIGHT NOW AND TAKE HIM HOME TO HIS MOTHER."

32

We are awake, the baby and me, and on cable there's an HBO documentary about the Beslan elementary school siege. They'd interviewed some of the kids who made it out, giving them cameras to take around the school: empty, blackened, untouched for four years. This is where they held us. This is where they shot my mother. This is where we drank pee 'cause there was no water, and it was so hot, and so many bodies crammed together, and everyone was scared. One of them talked about how sad her town was now. Everyone wears black, there's no dancing, no laughing, and she wants to grow up and leave and find a place where it's okay to be happy.

32

We got rid of the fucking cable.

32

We are awake, the baby and me, and I call my mom crying. I'm so tired, so scared, so in the fog of it, and after a brief conversation, I pack him in the car seat and drive the four hours overnight to Chelsea. You need a place where you can rest.

You need to let yourself crack. You need someone to take care of you.

For me, that means my mom.

How long did I sleep? Minutes? Days? When I finally wake up, it's the middle of another night and everyone else is asleep. I remember getting into the car, turning on the brights, and driving west from downtown to where the train tracks cut across the road. I follow them till they turned into a slingshot, park my car, and walk down to the shed, amazed and relieved that it was still there. I sit on the little steps and look up at the stars—when was the last time I saw them? My entire adult life, I'd lived in cities: Boston, Florence, Chicago, Prague. You can't see the stars from the city, at least not like a blanket above you, or the inside of a telescope, or the answer to an impossible question. My time in Chelsea is the only time I've had them at the ready, just a tiptoe down the stairs and out the back door, careful that it won't slam behind you and wake up your parents. I think of those midnight bike rides down Cavanaugh Lake Road, fourteen, fifteen, sixteen years old and all I wanted to do was scream. Now I was thirty-two, sitting in the darkness, trying to see my own hand in front of my face. Thirty-two and waiting for the train. Waiting to throw my voice at it, the exhaustion and the fear.

32

At a routine postpartum checkup, the doctor tells me there's a cyst on my ovary. She makes a grabbing gesture, wrapping her fingers around something unseen and ending in a fist. "Just cut it out of me," I say. I am not thinking clearly. I am not thinking at all. The baby still isn't eating well and I am still not sleeping and it's dark in here. When I think of the word "here" I make a gesture toward my brain that nobody understands but me.

We talk options. There's a chance I could lose the ovary and immediately I'm thinking: *Wait, what about another kid?* Do I want another kid? Where would I put another kid? Who would hold another kid? How would I feed another kid? I can't feed the first kid. "The medical odds of getting pregnant are fine with one ovary," my doctor assures me. "We just need to be realistic." I laugh out loud. Realism is not my strong suit.

I'm not allowed to eat or drink anything the night before surgery. My body needs to be TOTALLY clear of fluids; I am given a pamphlet and the word TOTALLY is in all caps. By the time I get to the hospital I am so hungry I want to eat my own arm. We get to pre-op, fill out paperwork, and ask for the bathroom to make sure I'm totally TOTALLY clear. After that, I'm led to a prep room, handed the backless robe, and a very nice woman named Lois comes in with a plastic cup and tells me she needs a urine sample.

"I can't," I tell her.

"We just need a little."

"I don't have a little."

"A few drops. You can manage a few drops, right?"

I imagine I am the warrior Yu Shu Lien from *Crouching Tiger, Hidden Dragon* and I cut off Lois's head with the Sword of Destiny. Then I go back to the bathroom. What follows is ten ridiculous minutes trying to conjure something out of nothing while Christopher gets out his laptop to "live tweet" this whole experience on a new social media platform called Twitter, which ended up being great because it cut the tension and proved helpful in later writing this account. Seemingly endless anesthesiologists come in, all saying the same things: What diseases are in the family—*heart, skin*. Are you allergic to anything—*no*. You'll feel a burn when I put in this IV—*yippee!* That's an interesting tattoo. What does it mean? And while all of this is happening, my doctor draws on my stomach in purple marker and Lois tests my measly pee in case I'm pregnant.

"I *just* was pregnant." I tell her. "I *just* had a baby."

"Are you breast-feeding?" she asks.

I am silly from the IV. *Tee-hee* BOOBS.

"Mama's gonna have moonshine in her milk tonight!" somebody says. I remember laughing: at the moonshine, at Lois, at the anesthesiologists, at the paperwork. *Things might go wrong. Sign this form confirming you've been told things might go wrong.* "The surgery will take two hours," my doctor explains. "Afterward, you can go home once you walk across the post-op room."

"How long will that take?" Christopher says.

"Four or five hours," the doctor says.

"This will feel like four or five martinis," the anesthesiologist says.

I must have passed out because the next thing I remember is waking up. My first thought: *I don't want to be here anymore.* So I get up and walk across the room.

"That was fast!" says the recovery nurse.

"Thanks!" I say. "I'm going to puke!"

Christopher comes in with crackers and juice and Norco, and explains that I still have both ovaries; the doctor showed him pictures. I remember feeling relieved. I remember feeling fuzzy. I remember getting home that night and my friend Amanda was there with my son. She handed him to me and I cried, not because I'd been scared something would go wrong during the surgery but because surgery is a scary thing, period, never mind when there's a new little person needing me to wake up the next morning and feed him and love him and teach him that sometimes—most of the time, all of the time—we are tested.

32

I NEED HELP.

I stared at those words in my journal for weeks. Eventually I decided that HELP was too big, too much, too terrifying. I crossed it out and wrote A SHOWER instead.

I could do that. I could get up off the floor and take a shower.

Remember those Herbal Essences commercials in the nineties where the woman has an orgasm while shampooing? It was like that.

The next day, I wrote: I NEED TO MAKE THE BED.

I made the bed and wrote: I MADE THE BED!!

Two exclamation points, that's how fucking huge this was.

Next: I NEED TO WALK THE DOG.

I walked the dog and wrote: I WALKED THE DOG!!

Next: GROCERY SHOPPING.

I WENT GROCERY SHOPPING!!

COFFEE.

I GOT COFFEE!!

I WROTE A POEM!! A SHITTY POEM!! BUT I WROTE IT!!

I SANG TO THE BABY!! HE LIKES BIG STAR!!

I GOT THE BABY IN THE BACKPACK THING!! I DIDN'T DROP HIM!!

WE WENT TO THE MONTROSE DOG BEACH!! MOJO IS SO HAPPY!!

THERE'S SOMETHING WEIRD WITH THE BABY MONITOR I CAN SEE OUR NEIGHBOR'S BABY ON THE SCREEN WTF??

I WENT TO BUY PANTS!!

FUCK PANTS!!

WE DID NOT WATCH THE FINALE OF THE WIRE.

WE DID NOT WATCH THE 2008 PRESIDENTIAL DEBATES.

WE SLEPT.

WE SLEPT.

WE SLEPT.

Every day, I made lists of these seemingly small accomplishments until one day—and maybe this sounds ridiculous but I swear it's the truth, a memory so tangible I can hold it in my hands—I walked into my son's room and I saw him. He was beautiful and perfect and laughing and I *saw* him. I could live that moment from the movies where the mother holds her child and her heart cracks open and how can you breathe under the weight of all that love?

33

Barack Obama won the election and Chicago was a dance party. You could feel it: walking down the street, riding the L, strangers high-fiving. For months, I'd been living in a bubble: my baby, my job, everything inward. Stepping back into the world at this particular moment in history was a gift.

I want to remember our country's capacity for joy.

He gave his acceptance speech in front of 240,000 people at

the exact spot of the 1968 Democratic National Convention. "History gave Grant Park another chance," said CNN, "as the scene of a peaceful and jubilant celebration." I watched the setup from my office windows in the South Loop, the crowd filling the park. Then I headed home to watch with my family—blood and chosen.

33

I bent over the bathtub to pick up my son and my back snapped. Both of us hit the floor. He sat up and laughed, naked and dripping. His laugh was the most glorious thing I'd ever heard.

I tried to sit up and discovered I couldn't.

33

I was scared of the physical therapist. I might hurt myself again.

33

I was scared of the gym. I might hurt myself again.

34

I was scared of the yoga mat. I might hurt myself again.

34

I was scared to be touched. I might hurt myself.

35

I was scared to sleep. I might turn the wrong way in my sleep and hurt myself again.

35

I was scared of shoes. The arches could hurt me.

35

Christopher gave me an album called *Into the Trees*, by Zoë Keating. She's a one-woman orchestra, a classical cellist who uses live sampling to build an entire planet of sound. One of the songs in particular cut into my heart. It's called "Optimist," written for her unborn son.

What's the song that saved you?

In the many moons ago of pre-Internet streaming, I'd buy a CD for a single track. *Nothing's Shocking* for "Summertime Rolls," *Blue* for "Case of You," *Exile in Guyville* for "Fuck and Run," which I'd play on repeat ad nauseam until (a) I scratched the disc, (b) I lost the disc, or (c) whoever I was living with "accidentally" scratched and/or lost it, which honestly, I get. Other people's obsessions are infuriating. My husband—a man who stood on a beach and promised to love me forever despite my obsessions—recently took to social media for support: "Megan has listened to the same song nine hundred times in a row," igniting several discussion threads including best practices for couples who work from home, superior brands of noise-canceling headphones, and "omg when

I lived with her in [year] she did the same thing with [song] and now I can't listen to [band]."

This originated, I think, with Paul Simon. My dad and I logged countless hours driving between the town where we lived and the town where he worked, rewinding *Graceland* and singing our faces off. Long before I knew what a cinematographer was, I knew: "Don't I know you from the cinematographer's party." I knew: "Aren't you the woman who was recently given a Fulbright." I knew: "Who am I to blow against the wind," my hands making snakes out the window on M-52.

"Girls Just Want to Have Fun," rewinding till the tape tangled. "Everybody Knows" from the *Pump Up the Volume* soundtrack. "Me—Jane," "Walk Away," and this song by Muki that goes: "And I don't want to know about evil/ Only want to know about love." When my son was born we listened to "Thirteen" nonstop. It was the only thing that would get him to sleep. For new parents right now in the thick of it—I see you, your incomparable love and incomparable exhaustion and beautiful, terrified hearts—maybe give it a try? Big Star, off *#1 Record*. iTunes tells me I played it just under ten thousand times, the second-most-played track in my library.

The first was "Optimist." It's all I listened to that summer, my unofficial theme song. Rocky had "Eye of the Tiger." Judy Bernly had "Nine to Five." Kira had "Magic." The Kid had "Purple Rain." And Rachel Marron had "If I should stay/ I would only be in your way." But "Optimist"—that was for me. I listened to it on the L, back and forth to my day job in faculty development and my night job teaching writing. I listened to it at the park, chasing my then two-year-old son.

I listened to it walking Mojo to the Montrose Dog Beach. I listened through the winter, gray and drudge and ice; on my way to doctors' appointments, trying to figure out what was wrong; and in the Uptown condo we couldn't sell.

It gave me back a part of myself I hadn't known was missing.

36

Caleb is very sure he's Superman, which sounds cute except we live on the third floor and he keeps trying to fly. Last week he stood at the sliding balcony door in his red-and-blue costume, his nose pressed to the glass. "Mommy," he said. "Let me out. I'll put my arms out far; I'll go high up in the sky."

36

I spent two weeks at the Ragdale Foundation, an artist residency in Lake Forest. It was my first substantial length of time away from my son. I had plans that were so high-stakes, so high pressure: finish a draft of my book, get back into a day-to-day writing process, and something that I referred to as "conceptualize next project," a phrase that nearly sent me into a panic attack whenever I looked at my to-do list.

Know what I did that first day at the residency? Slept. For, like, eighteen hours straight. Then I binge watched season three of *True Blood* on my laptop—twelve episodes without stopping. Then I cried for the time I'd wasted. Then I cried because I missed my kid. Then I cried because Tara was mad at Sookie. In the end, I scrapped my plans, went for walks, stared at the wall, read a ton, and finished one short essay about postpartum depression.

In every room at the residency, there are spiral notebooks where artists can document their time. On my last day, as I was packing to leave and berating myself for how little I'd accomplished, I flipped through that notebook, reading wisdom and advice from writers who'd worked there before me. Thrillingly, I found a name I not only recognized, but emulated: a poet and playwright named Coya Paz. She'd been there, in that very room! Sleeping and reading and creating, just a few months before! I dug into her words, expecting to find something about the projects she'd completed, or maybe some profound musings on the artistic process.

Know what she wrote about instead? Sleeping. Staring at the wall. And—I shit you not—watching *True Blood* on her laptop, and what was up with Sookie?

I shy away from giving advice to writers and to parents. We have different situations, different processes, different challenges and expectations. That said, I think what I learned at that residency might apply to all of us: Be gentle with yourself. The writing process is more than building sentences.

36

There's an envelope taped to our bathroom mirror. A reminder, if you will. It's addressed to the bank and, for now, it's empty, but when things get really hard, we joke about mailing back our house keys and throwing in the towel. We'd been having that conversation a lot. Our home was worth less than half of what we paid, one the 11 million in the United States defined as underwater. Three years trying to sell, six jobs between the two of us.

Here's the truth: we were lucky. We knew where our next

meal was coming from. Knock on wood, if one of us got hurt, we had insurance, at least for the moment, though who knows if we could afford the deductible.

The question of walking away from our mortgage was about how we were going to live.

Not if.

Here's a metaphor: a few years ago, my dad got attacked by a bear. Imagine it: the average Kodiak brown bear weighs fifteen hundred pounds. They are five feet tall with all four paws on the ground, and over ten feet standing on their back legs, which is what that bear did when she saw my dad. Look up: ten feet is as tall as the ceiling. The bear is as tall as the ceiling. She is scared, she is pissed, and there, in front of her, is my dad and she goes for him: fifty yards, thirty yards, ten—one swipe of her claws and my father might cease to exist completely. For years, he has prepared for this moment, pulling his rifle on rabbits, partridge, deer, caribou, all requiring singular, focused aim and the bear hits the ground with a bullet in her brain.

On the worst days I wondered if it was easier being attacked by a bear. A nearly impossible shot, sure, but at least you know your target: here, between the eyes. Right now, in this time, this economy? I don't know where to shoot. I don't know who to blame. There are numbers and statistics and layoffs and something called, fittingly, bear markets. Google tells me they indicate a general decline in the stock market: "The transition from optimism to widespread fear."

That fall I came across a literary website called *The Rumpus* and, in it, a weekly advice column by an anonymous writer named Sugar (Cheryl Strayed). In a letter called "The Future Has an Ancient Heart," I found this:

> You don't have to maintain an impeccable credit score. Anyone who expects you to do [that] has no sense of history or economics or science or the arts. You have to pay your own electric bill. You have to be kind. You have to give it all you got. You have to find people who love you truly and love them back with the same truth. But that's all.

I went into the bathroom and pulled the envelope off the bathroom mirror. Then I went to the apartment search on Craigslist and typed in the word *beach*.

37

New neighborhood, new apartment: five miles north to Rogers Park where the streets dead-end into community beaches along Lake Michigan. Every morning I walk my dog down Tobey Prinz Beach, a block of sand at Pratt Boulevard named for the local activist who saved it from real estate developers in the 1950s. On Sundays, church bells ring at the Madonna della Strada Chapel a mile down the lakefront. "They sing in celebration of your memories," the Internet tells me, "of your achievements, struggles, and hopes."

Here, we're climbing out of the mess.

Less screwed, less scared, still climbing.

37

Christopher wakes me up in the middle of the night to show me an e-mail from a celebrity whom we shall not name. He wants to tweet about the blog, but his retweets sometimes crash websites so he likes to ask permission first.

I want to hug this celebrity.

When he tweets about the site, traffic spikes. Our rent gets paid. Followers multiply. What was once a side project is now a possibility.

It's made me look very closely at how we use our platforms, whatever the size. The seemingly smallest gestures can mean the world to someone else.

37

December 14, 2012. Twenty children between the ages of six and seven and six grown-ups were shot and killed at Sandy Hook Elementary School. I sat in front of my laptop, watching Twitter—can't move, can't cry, breath locked. Every time there's a mass shooting, I spiral—Columbine, Virginia Tech, Aurora—and I'm back to that day in 1993. Sometimes I imagine my dad. Sometimes myself. Sandy Hook was the first time I imagined my son.

Not long after, Stephen Leith—currently serving a life sentence for first-degree murder—was interviewed about his opinions on gun violence. "I wouldn't care if there was a background check on me," he told the reporter. "I was a respected science teacher. I was a law-abiding citizen." In other words, a good guy. A good guy with a gun—eleven guns, including the AK-47. "I figured at one time they would not be allowed to be sold, and I wanted one in my collection."

I called my dad and told him about the interview.

"Fuck him," he said.

It's the only time I've heard him use that word.

37

"I'm scared of writing."

A student says this to me. I ask her why and she says, "Because __________ might read it." When you teach writing, you hear this a lot. The impulse is to say: *Who gives a shit about* __________. But it's more complicated than that; in fact, I think that response is dangerous. While _________ might be the writer's mother who doesn't know the writer smoked pot that one time, I've also worked with writers coming out of abusive relationships; writers interrogating identity, race, gender, and sexual orientation; writers surviving physical and emotional violence; writers surviving physical and mental illness; writers not just surviving this life but living the holy hell out of it. The question of who's on the other end of the page is more complex once I look past my own privilege. It's an issue of physical safety. Above all else, I want the artists I work with, specifically youth, to be safe. That's when we can ask hard questions, face difficult truths, make discoveries and go deeper and push and fight and learn through the work, and I think part of that includes a differentiation between the act of writing and the choice of when and if and how to share that writing. Maybe, for now, you're just writing for yourself. Maybe you're writing for a teacher. Maybe, if your teachers do their jobs right and build spaces where art and ideas and individuals are all equally respected, you'll choose to share it with the class. Maybe you'll choose to perform it. To submit it. To hand it to ___________ and start a dialogue, or fuck ________________ and give it to the world.

37

Dear Stephen Hnatow My White Knight of a Realtor Who Swept In at the Last Possible Second and Sold My Stupid Fucking Condo as a Short Sale So We Didn't Have to Foreclose Even Though We Lost a Ton of Money and Will Be Building Our Credit Back Forever But Who Cares We're Freeeeeeeeeee,

I love you.

If I'm ever ready to consider the possibility of having a discussion about maybe talking about buying something again in the far-off distant future, I will call you.

37

With the mortgage off our backs, my husband quits his day job to blog full-time.

37

With the mortgage off our backs, I write.

37

With the mortgage off our backs, I go to the doctor, a woman who is smart and kind and has great shoes. We talk about heart disease. We talk about melanoma. We talk about my thyroid, which, she discovers, doesn't function. She looks at my THS levels and asks about weight gain—*yes, lots.* Dry skin—*yeah, but I live in Chicago.* Irregular/difficult menstrual cycle—*I did, it was awful, but I got an IUD and now it's fine and omg what a luxury, to walk through this world without pain!* Depression—*I'm better now!* Memory loss—*I'm writing a memoir! Or at least a memoir-like thing? Genre distinctions are confusing and who knows what the marketing department will suggest we call this book in 2017.* And exhaustion—*maybe?*

"Maybe?" she repeats.

I say I have four jobs.

I say I have a five-year-old kid.

I say I get up at 5:00 a.m. to write.

I say, "Exhausted? Fuck—aren't we all?"

37

My husband calls and ask if I can sneak out of work. We do this occasionally. Cubicles are dull and sex is fun, but this time, when I arrive at the address he's given me, it's not a public restroom or a freight elevator. It's a car dealership. We're not here for a quickie. We're here to buy a car, which to some of you may sound easy and normal but for us was a big goddamn deal. For years we've been driving a beater of a Honda Civic: it started maybe seventy percent of the time, made weird noises one hundred percent of the time, the radio never worked, the AC never worked, the windows didn't roll down, Christopher is six foot five and sick to death of scrunching. But finally, with the mortgage off our backs, we can buy something—not *new*, but new for us.

Afterward, we pick up our kid, now four, from preschool, and are driving somewhere or other when we hear, from the backseat, "Daddy, do we have Xfinity?"

Christopher says, "No, buddy," and our kid loses his mind,

zero to fury within seconds. "WE HAVE TO HAVE XFINITY XFINITY IS THE BEST XFINITY XFINITY XFINITY."

"What on Earth?" I say, turning in my seat.

He points forward, between the passenger and driver seats, to the radio. "The man in there said so!" he says, his panic real, and with the same type of *screeeech* you hear in race-car movies—say, *Death Race 2000* or *The Fast and the Furious* franchise—my husband crosses Western Avenue, pulls the not new car off to the side, and turns to face this American child of 2013 who watches Apple TV, listens to podcasts and playlists, and until this very second has never heard a commercial.

"We need to talk about something important," Christopher says, and our son goes focused and quiet. Daddy's voice is serious. Daddy uses this voice to talk about bullying and typography, *say please and thank you,* and don't kick the dog. "We need to talk about the psychological impacts of deceptive marketing."

He's seven years old now and recently, as I drove him to karate, he leaned forward between the passenger and driver seats and yelled: "THAT IS A LIE. AMERICA DOES NOT RUN ON DUNKIN'."

37

I told a story at a local reading series about sleeping with my TA in college. Afterward, an editor from a new independent press approached me and asked if I'd like to make a book.

37

I was sitting outside my son's kindergarten class, waiting for him to put his rain boots on and I got an e-mail saying that the piece I wrote about postpartum depression would be included in *The Best American Essays.*

37
Tentatively, carefully, cautiously—things were looking up.

38
Our building caught on fire.

38
Christopher took me to hear Zoë Keating perform at the Old Town School of Folk Music. If you have the opportunity to hear her live, run. Do not walk. Sitting in her audience, watching her build "Optimist" from a single refrain into an entire orchestra, I knew I wanted to make something like that.

A week later, I jumped in on a story development session at 2nd Story, the storytelling collective I'd worked with for nearly a decade. Picture twenty-some brilliant, complicated people telling stories informally over coffee and/or wine. You can hear the ones that take hold, that stick, that get people leaning forward and asking questions. Those are the ones we develop: first on the page with other storytellers and an editor, then off the page with a theater director and sound designers. And then we share them with a live audience, which, depending on the venue, might be eighty or a hundred or five hundred of your new best friends.

Amanda was facilitating that day and she asked us all to stand. "Imagine that wall is the day you were born," she said, pointing. "And that wall is, say, age forty." She pointed at the opposite end of the room and indicated the empty space between: ten years old, twenty years old, thirty. Then she invited us to stand on a time when we were afraid.

I was nineteen years old and alone in a hostel in Rome, sixteen and naked in a quarry in Michigan, thirty-six on the phone with my dad, thirty-six on the phone with the bank, twenty-two on a side street in Wrigleyville, eighteen in my dorm room in Boston, thirty-two on the Montrose Dog Beach, thirty-two on my bathroom floor while my baby cried on the other side of the door, thirty-eight in the car with my son on my lap, our building lit up red with sirens.

Do you want to play, too? You can do this on your own: grab a sheet of paper and draw a horizontal line between two X's. Do it. I'll wait. The X on the left? That's when you were born. The one on the right? That's however old you are right now. Take thirty seconds and mark some X's on that line for the moments that scare you; the big and the small, the wonderful and the awful, when you were six and twelve and twenty and forty. Don't think too hard about it; just get it out of you. See what you have to say. I wager you'll find some beginnings there, some meat and emotion and story. Write them. Paint them, dance them, scream—make something.

38

Caleb and I make a deal: when we watch movies, the remote control will be *right here.* At any point—if we're scared or sad or confused—either one of us can hit Pause. We can talk about it. We can fast-forward or turn it off and play Uno.

Examples: I paused *Honey, I Shrunk the Kids* because I'd

forgotten how sad it is when the ant dies. "It's okay, Mommy," he said, patting my leg. "It's not real." He paused *Return of the Jedi* to ask when we could see Leia's movie. "We've watched three movies about Luke learning the Force. Leia's got the Force, too, right?"* I paused *The NeverEnding Story* right before Artax dies in the Swamp of Sadness.

"Listen," I said. "Something scary is about to happen."

"How do you know?"

"I've seen this movie one or two or nine hundred times."

"Does this part scare you?"

"It does."

"It's the horse, isn't it?"

"It is."

"It dies, doesn't it?"

"It does."

"Okay. Let's fast-forward."

The next day, he asked if we could watch it again. Chicago was mid–polar vortex and all the schools were closed. Here is my privilege: I could work from home. We sat on the couch, Caleb watching the movie while I answered e-mails, looking up in time to hit Pause at the Swamp of Sadness. "We're at the horse part," I told him.

"I know," he said.

"What do you think?" I said.

And he said, "Let's try."

We made it fifteen seconds.

Jump to the next day, round three. I had my hand on the remote the whole time. "Not yet," he'd say. "Wait." We made it through the scene. Then we hit Stop and talked for a while.

---

* That's my boy.

In the end, we both cried for Artax, for Atreyu, for Bastian. "It's okay if a story makes you sad," I told him. "It's okay if it makes you angry or afraid. These feelings are real. Let's live them."

39

My dad calls from Alaska. He calls often, a couple times a week. He tells me about the weather or my brothers or whatever book he's reading, and even though it's always fine, I still feel the smallest kick of panic at his name on the caller ID.

Is it the mountain?

Is it his heart?

"So listen to what happened," he said. "I was chopping stuff for stew and there's a knock on the door and I answer it and it's a guy selling home security systems. Door-to-door home security systems. I haven't seen a door-to-door salesman in I don't know how long. You know I was a door-to-door salesman, right?" I didn't. "I told you that, right?" He hadn't. "Dictionaries. I was in college and I drove to Pittsburgh and sold dictionaries."

And we're off, me with the questions: "Wait, you lived in Pittsburgh?" and him in the story: "While I was there I saw Anna Maria Alberghetti play Maria in *West Side Story*. It was the first time I'd seen theater on that scale and I was in awe. There was this chain-link fence across the stage and when the Sharks and the Jets did that *ba-da, ba-da-da, ba-da-da-da-da* number they jumped on the fence and it didn't move! It didn't even bend! I was shocked! How did they do that?" I love listening to him. I love that he talks about *West Side Story* at the Civic Arena with the same excitement as he talks about a twelve-point buck or a hundred-pound halibut or a whale swimming under his boat, so close that waves start rocking and you have to grab hold of something so you won't fall overboard.

"What we were talking about?" Dad says.

"Uhm—Sharks, Jets, Alberghetti—"

"Pittsburgh, dictionaries, door-to-door—"

"Security systems."

"Security systems! So I was cutting stuff for stew . . ." and off into the story. Apparently, the salesman pushed hard. You and your family are at risk. Here are the statistics. Your dog can't protect you. Your guns can't protect you. And forgive me for saying so, sir, but you're not a young man anymore. There are assholes out there who'd take advantage. They'd come in when you're sleeping, when you're out of town and your wife's all alone and—

"You know what this guy was doing?" Dad says.

He pauses.

He's a great storyteller—building the tension, landing the punch.

"Selling fear."

39

A local theater company invited me to be part of the post-show discussion. A character in the play had postpartum depression and they were interested in how art can help us heal. I loved the production—the director and cast both brilliant. The theater itself was small, black box, with actors and audience at the same level. The lights went down, all of us together in the dark. There was the new baby. The confusion, the fear.

The mother was crying center stage and there I was on the bathroom floor—can't move, can't cry, breath locked. It was so real. She was right there next to me. I could reach out and touch her as she moved through the fog. Another character entered: a teacher trying and failing to connect. Another: a teenage girl all up in her head, the real world around her dull and on mute. Three different woman, a triangle of bodies, their stories tangled together and there, in the dark, I lived all of it: my body, my breath, my bones.

Afterward, there was a panel including the director, a social psychologist, and me.

We talked about the data: six hundred thousand women experience postpartum depression annually. We talked about the nuance of that data: the number goes up to nine hundred thousand if you include miscarriages and pregnancies not brought to term, and that's based only on reported diagnoses. Here's the truth: only fifteen percent of women with postpartum seek help, and that percentage changes based on class and race, postpartum as experienced by adoptive mothers, and queer and trans parents.

We talked about the feeling—guilt, shame, helplessness—and how we carry it all in our bodies. I remember saying how scared I still was seven years later, that it would all come back: one second you're fine; the next, you're on the floor. And we talked about the stigma, how so many women don't seek help because they don't think people will believe them.

A woman in the audience raised her hand. She said she didn't believe them. Postpartum depression wasn't real. Women today—

That meant me.

—just needed to love their babies. To realize that a baby is a gift. Another audience member agreed: Maybe, she said, if those women—

That meant me.

—stayed home with their children, they wouldn't feel so guilty.

She went on, but I'd stopped listening.

I will permit it to pass over me and through me.
Where the fear has gone there will be nothing.
Only I will remain.

Here's how I've decided to remember this experience—the real estate I give it in my memory. This was a small taste of what it feels like to be invisible. Multiply it times a million and you've got what others live through every day. Like their lives don't matter. Their voices don't matter. Their stories don't matter. Their gender doesn't matter. They don't matter because of the color of their skin or where they live or their faith or their job or what's in their bank account or whether or not they have children or who they love or how old they are or how their bodies move or any of a thousand things.

I'm grateful for the empathy.

Dammit, I'll use it.

39

I'm at a writers' festival in Florida: heat, sun, ocean. My body knows lakes and rivers so salt's always a shock. I walked in burning sand and drank whiskey and talked about stories with smart people hungry for ideas, for words. I love these people. They light a fire under my ass and I do my best to return the favor.

On the third day, I got onstage and told a story. It was about a girl, nineteen years old, scared and alone in a bathroom stall in a hostel in Italy. She pees on the stick. She waits three minutes.

I tell this story often, so much so that it's no longer mine.

At the reception following the reading—drink wine, eat cheese, talk to the writer—a woman approaches me, asking if I have a moment. We go off to a corner. She tells me about a time when she was young and scared and alone. I'm not going to say it here. It's not mine to share, not to mention this woman is an incredible storyteller. If and when she's ready to put her work into the world, it'll crack open your heart.

I will say this much: It was heavy. Guilt and shame and fear.

Way too much for a person to carry alone.

I held out my hands and asked if I could hold it.

Real and Imaginary Ghosts

On Sundays we go to the Montrose Dog Beach so Mojo can run it out. It's a mile of fenced-in off-leash shoreline on Lake Michigan—open to the water, open to the sky. If you show up midmorning—say, 10:00 a.m.—it's a madhouse. It's a jam-packed outdoor party with hundreds of dogs playing, rolling, chasing balls, chasing each other, sniffing and/or mounting, saying no to the sniffing and/or mounting, which occasionally results in loud scary baring of teeth at which point their humans jump in and we all have a little lesson on being part of a community—which frankly I think is more necessary for those of us with two legs than those of them with four—and then back to the running and the splashing and the grand ol' goddamn time.

Up front: I am a person who loves dogs.

That said: shout-out to the cat people who think a dog beach is hell on earth. I see you. You're still welcome in this essay, which is not at all about dogs and yet totally is.

*

One night—this feels like a lifetime ago—I went early to Montrose, 2:00 a.m., maybe 3:00. I'd been up feeding the baby and couldn't get back to sleep, didn't sleep much then, a dead zombie fog. Our condo was teeny tiny, all of us on top of each other, and I was scared that if I turned on the TV or turned the page of a book or moved too fast or breathed too loud or thought too hard I'd wake them: my perfect, months-old little boy *please go to sleep baby, please please please* and his exhausted dad, working two full-time jobs to keep us afloat. Mojo and I tiptoed through the dark hallway then out the door, down three flights of stairs, and into the street-lit city life of late-night Uptown, bodies lined up for shows at the Aragon across the street and the Riv down the block and the Green Mill around the corner, everyone drinking and dancing and laughing and partying and Lord knows what other wonderful and terrible things. It was too much, too bright, too loud, too alive. So instead we headed east—*are you crazy?*—down Lawrence. *Walking that street?* Under the overpass *in the middle of the night?* Through the park *alone?* Over the hill—*fucking stupid*—and onto to the dog beach. It was early summer. The lakefront was empty. Gloriously, breathtakingly empty. Wide and open and empty. And quiet. So quiet. I could hear my own heartbeat.

Shhhhhhhh—listen.

Wind: that night, gentle. Waves: that night, calm. The city sounds far away behind us. My feet sunk in sand, sand between my toes, sand kicked up and flying as Mojo shot toward the water, so much energy kept under lock in our tiny condo with its crying baby and his crying mother—*shhhhhhhh.* A stripe of light ran from the edge of the water to the moon. A path, I thought. Away from all of this and up into the sky.

*

I found Mojo on Petfinder, an online service connecting adoptable animals with their human families, like Match.com but without the algorithms and gross creepy dudes. Are you in the market for a dog? A cat? A ferret, a pig, a Burmese python? This is the place.

"Christopher, look at this dog," I said. Christopher was my boyfriend. We'd just returned to Chicago after a year in Prague, me teaching in an American study abroad program, him designing websites for an American advertising company. Dogs were everywhere in Prague: the bridge, the metro, Kavarna Meduza where I wrote every day. Even the server had a dog who came with him to work, a sleek black lab, impeccably trained. "What can I get you?" he'd ask in Czech, his dog sitting next to him, still as stone. I'd order—coffee or wine, depending on the hour—and he'd walk behind the bar, his dog following right behind, and return with my drink, dog at his heels. I wanted that dog. I wanted *a* dog. I wanted *our* dog, and so, like many recently shacked-up couples, we decided to adopt a puppy. "We'll name him Egon," I said, thinking of the Schiele Art Centrum in Cesky Krumlov. But that seemed unbearably pretentious so I switched the reference to Harold Ramis, aka Spengler in *Ghostbusters*. After a little research, we found out that an animal from overseas meant months in quarantine after returning to the States, so we agreed to wait. For the rest of the year I walked around singing, "Doe—Ray—Egon!" like the Ghostbusters do when they turn on their proton packs.*

---

* Now I lick imaginary lasers like Kate McKinnon aka Holtzmann in the reboot.

We'd been back in Chicago for a week when I started casing Petfinder, even though the timing was impossible. It was February, another awful winter. We were working too much; me still teaching, Christopher still in design. We had to make money; I picked up weekend shifts at the brunch place where I'd worked in college. We had to find an apartment; my friend Jeff let us crash with him but it was too tight for three. We had to kick the culture shock; a single trip to an American grocery store proved panic inducing. There were so many products, so many options. Do you know how many brands of cereal there are in the United States? It's too much cereal. Who the hell needs that much cereal *cereal cereal cereal*.

A puppy needed time we didn't have. We'd wait—the summer, maybe, when everything was, if not easier, at least sunnier.

But here's the thing: There's never a right time to have a dog. To have a child if you so choose. To fall in love or write a book or perform a poem or put your heart on the goddamn table. *I'll go to the next open mic,* we say. And: *Next month I'll sign up for a class.* And: *I'll quit my job later.* Doesn't matter that I'm miserable, I have to wait for the lateral move or the substantive raise or the fully funded position and hey, how's that going for you?

You wake up one day and there it is. You say yes or you say no.

"Christopher, look at this dog."

His name was Mojo. They weren't sure what kind he was, but for sure part pit. He was six weeks old and so adorable I thought I'd die. Red brown, with nine black toenails and one white one, on his right front paw. Christopher downloaded his picture. He photoshopped a text bubble over his little puppy head. It said: WILL YOU BE MY MOMMY?

*

I am not proud of what happened next so I'll say it fast: When I got to the shelter the next morning there was a little girl, eight years old, maybe. She pointed at Mojo and squealed, saying, "Daddy, look at that puppy! Can I have that puppy?"

"That's my dog," I said, and I shoved her out of the way.

*

Dog people are obnoxious. We only talk about our dogs. If you met me around the time I adopted Mojo, here's what I would've told you:

Me: Mojo is afraid of stairs! Mojo ate a lightbulb! Mojo smells like toast! Mojo was the valedictorian of his dog training class because he's totally the smartest or maybe because we carried hot dogs in our pockets or maybe because the other dogs in the class were super expensive pedigreed inbred purebreds, Weimaraners and puggles and something called a doodle who, his owner informed us, *was descended from kings*. And Mojo and I looked at each other like: *The fuck is this guy?* while the doodle pooped on his own food. Every morning we go on walks in Humboldt Park, and we pass all these dogs, and always, about eight feet apart, their owners and I wind our leashes around our wrists. "It's okay," I call out. "He's friendly!" "So's he," they call back, and we relax our grips, and our dogs sniff each other with waggy tails.

It's got me thinking: Wouldn't it be great if it could go like that with people? You get to a certain distance and call out: "It's okay! I'm friendly!" or "Steer clear, I'm an asshole!" or

"NO MOTHERFUCKER I DID NOT HURT MYSELF ON MY FALL FROM HEAVEN."

One time—I'd been in London for a conference so I was crazy jet-lagged and Christopher had been up all night at the Puerto Rican People's Parade, i.e. the best party in the city, point being we were exhausted, sleeping like bricks—and we wake up to Mojo peeing all over the bedroom 'cause we'd forgot to take him outside. WE ARE THE WORST—and also the power was out for some reason I can't remember. Anyway, Christopher's trying to find a flashlight and he reaches into the Box of Random Stuff and accidentally grabs a food processor blade so now he's bleeding like a faucet and we go to the emergency room for stitches and Christopher is, uhm, not so great with blood and passes out on the gurney and the nurse asks me for his social security number. Show of hands: who's memorized their partner's social security number? I didn't know my own social security number, not till two years later when we bought a condo and signed a thousand documents and then another thousand later when we lost the condo, but regardless, the nurse was all sorts of snippy that I didn't know his information. "What *are* you to him?" she asked. I thought I'd black out from the rage: first at her, then at the American health care system, then at systems in general. I thought of my gay and lesbian friends who deal regularly with such bullshit as they try to take care of their loved ones. I thought of all the women I know who have been talked down to by medical professionals, made to feel stupid and small, like we don't know our own bodies. Also: Christopher had lost a lot of blood and I was terrified. Also: it was 2:00 a.m. and the fluorescent lights were terrible. Also: I was mad at the English language for limiting personal relationships to words easily understood by this nurse and her forms and their establishment, words like "boyfriend" and "spouse" when clearly there were levels in between like WE HAVE A DOG TOGETHER and while I was swimming in all that whatthefuck Christopher sat up on the gurney, white as a ghost, and said, "She's my domestic partner!" before passing right back out.

I knew then I was going to marry him.

I was covered in blood. I was covered in urine.

And fuck, I loved this man.

*

Fear is a learned response, so we taught the puppy not to be afraid. The Internet gave us a helpful and, uhm, thorough checklist of the people, places, and things to introduce to your dog prior to twelve weeks of age including children, escalators, men with beards, women in hats, and chickens. I lugged him around Chicago, checking off boxes. I like checking boxes. I like visible achievement: step by step, bird by bird. We went to the farmer's market. We went to the Wilson Skate Park. We drove in circles around O'Hare. Most everything on the list was easy to find: sirens, bicycles, people in helmets. I had the hardest time with carpet, of all things. Everyone we knew had hardwood: rickety, knotty, potential tetanus trap hardwood, yes, but hardwood nonetheless. Now, I would be able to ask on social media: *Who has carpet? Can my dog stand on it for a sec?* But this was the spring of 2005, a year after the launch of Facebook but a good two or three before we signed our lives away. (I think Facebook is Skynet. Take me with a grain of salt.) My bright idea: take the puppy to Empire Carpet! I'd watched the commercials for years (if you've never lived in the Chicagoland area, they play like hourly). Finally, a reason to go! I looked it up in the yellow pages and was surprised to discover there was no actual physical location; instead, someone is sent to your house with samples. I imagined the Empire Carpet Man showing up at my front door with squares of loop pile, cut pile, berber, and shag. Naturally, he'd bring The Fabulous 40s, the a cappella group that recorded the famous Empire Carpet jingle. Sing it with me, Chicago: "Five-eight-eight, two-three hundred, Empire!" They'd stand in my living room singing: "Up on the Roof," "The Lion Sleeps Tonight," and, my favorite, "Please."*

Instead of taking Mojo to Empire Carpet, I took him to another legendary Chicago institution: *Affordable Furniture & Carpet! Just Say Charge It! EZ Credit!* in Wicker Park. I set him down on a geometric area rug, he padded it with his tiny paws, and I checked "carpet" off the list, crumpled now at the bottom of my bag. We walked back out to Milwaukee Avenue and into a massive oncoming crowd, which I soon found out was a funeral processional honoring the death of

---

* In the absence of any knowledge of The Fabulous 40s, I apparently substituted the Nylons for this particular fantasy, a group that my high school choir friends and I adored. I'd forgotten about them until this very moment, pushed out of my memory from years of dating boys in indie rock bands, boys who scoffed at my love of PJ Harvey, boys who saw my copy of *Jagged Little Pill* and asked why the fuck was I listening to her, boys who would've most certainly ridiculed my love of a cappella. And if they didn't like my music, they wouldn't like me, right? Right? If there are any young women reading this and those above sentences sound familiar, if you're hiding parts of yourself to look cool or make someone love you, please repeat after me: fuck that noise. You are perfect. You matter. Hold on to what you love, the songs and books and style and obsessions and causes and questions that make you you. Find people who love these things, too. When you get lost, they'll help you find your way back to yourself.

Pope John Paul II. Traffic had been cleared and the street was packed, people walking calmly, serenely, some holding red-and-white flags, some holding portraits, some holding kids, holding hands, arm in arm, all quiet and singing softly. They kept coming, more and more and more. Later I read that an estimated ten thousand people were out that day, mostly Polish Americans from all over the Chicago area mourning the loss of the first Polish pope with a five-mile walk up Milwaukee from one prominent church to another.

I put Mojo down and we stepped into the wave northwest. At first, I did it to check boxes—puppy in a crowd, puppy in a parade—but within minutes it was something else entirely. Disclosure: my heart's not in organized religion. I grew up Methodist, Sunday mornings trying to sit still in the wooden pews and, afterward, cake on the lawn. I went to youth group every week in the church basement, but I can't tell you what was said. Something about being good. I remember we went camping in the summers near a river. There was a rope tied to a tree and we'd swing over the water like the Mountain Dew commercial, and later, sing around a bonfire. I remember how the music felt. I liked flipping through hymnals and joining in song, even if I didn't understand the lyrics. I liked the idea of bowing our heads together in prayer and maybe the force of our collective energy could cure sickness or slay dragons or move a mountain. I believe in things like that. Noah built the ark. The statue of Mary cries real tears. Gregor Samsa turned into a bug. A very old man with enormous wings made a leper's sores sprout sunflowers.

We could all use a little magic.

Where I get tripped up is when it's used to legislate. I dated a Young Republican for a hot second in college, a guy who called himself a Christian. He thought our country's problem was big government. I couldn't imagine a bigger government than one that would force me to have a child. We tried to understand each other, see eye to eye, but it never went well.

One night he yelled, "I DON'T WANT CONGRESS IN MY WALLET!"

"I DON'T WANT CONGRESS IN MY VAGINA!" I yelled back.

The air went out of the room. He stepped away from me slowly, as if I were a bomb. Was it the meaning of what I'd said that offended him? The fact that I raised my voice? That I used the word "vagina"? I think of him sometimes, like when that high school biology teacher was investigated for saying vagina in a class on the reproductive system, or when Michigan state representative Lisa Brown was told she'd violated decorum for saying vagina while discussing a bill on abortion regulations.

I wonder: where did he end up?

Please not the House. Please not the Senate. Please not anywhere that allows him to dictate policy over other people's bodies and other people's lives.

You know the last thing he said to me?

He said, "God help you."

Politics and religion tangle in my head in ways I can't see past, but on that day, that walk, all those people together in their prayer and their grief and their love—I got it. It felt like the storytelling workshops I ran in my friend Amanda's living room, a whole room feeling the same feeling. Recently I asked a friend, a woman very involved with her church, why it meant so much to her. She talked about the importance of community, of service, of giving back to this wild, beautiful world. I identified fiercely: what she found in church was what I'd found in art.

"But that isn't God," I said. "That's people."

And she said, "Same thing."

*

A few months after we eloped, Christopher and I bought a condo on Chicago's North Side.

On Sundays Christopher would grab coffee from the Starbucks on the corner, and if the weather was nice we'd sit in the sand. We talked about scary stuff—mortgage, bills, insurance. We talked about dream stuff—the book I wanted to write, the blog he wanted to build, the kid we wanted to have, and then, two years later, the kid we were having. Dogs kicked up sand around us. We stood in the water and threw tennis balls for Mojo. Children asked if they could pet him. I felt my own child kick and flip. Picture the time-passing montage from every movie ever: me on the beach, my stomach barely visible; me raking leaves, my stomach a soccer ball; me in a blizzard, my body its own planet.

*

New mothers are amazing. We're high as kites! If you met me around the time I had a baby, here's what I would've told you:

Me: The baby is cold! The baby is hot! The baby is hungry! The baby wants to fly! He doesn't like the book we're reading! He doesn't like his blanket! Somebody told me to get a Boppy and I wasn't sure what that was but the Internet told me and I went to Target and he had a meltdown in the checkout line and the woman behind me rolled her eyes. I try to be patient. Really I do, but hi there, excuse me, do you think you want him to stop crying more than I want him to stop crying? And hi there, excuse me, if you can't deal with children, don't shop where diapers are sold. And hi there, excuse me, I'm hanging on by a thread, I'm fragile as fuck, there are hormones coming out of my nose. Do you really want to take on a mother?

One night I saw a mother pushing a stroller outside after midnight and I was like: *That baby should be in bed. It's cold. It's late. What is that woman doing?* And then, a week after my son was born, my husband was in a car accident, nothing too bad but still. He called from the emergency room, he needed the insurance card, they were putting him in a cast, he couldn't drive. I got the baby into the snowsuit and the wrap thing and of course we couldn't take a cab—no car seat—so we got on the L downtown to the Loop, two o'clock in the morning, in February. I thought about that mother with her stroller at midnight and felt near blind with shame.

How dare I judge her.

How *dare* I.

There's a great essay about this by Aubrey Hirsch. She says judging another mother is like critiquing a woman being eaten by wolves. "That's not what I want to do," Hirch writes. "I want to say 'Hey, Mama! You looked like a badass bitch taking on those wolves!' and 'Aren't those wolves crazy?!' and 'Tell me how you're surviving these wolf attacks.'" That's the kind of mother I'd like to be: generous, empathetic, open to the new.

Yesterday I was listening to that Sleater-Kinney song, the one that goes "My baby loves me/ I'm so happy." I'd listened to it a ton before, but this time I was like: *Wait. What baby are we talking about? Baby like my lover or baby like my baby?* I hear music differently now. I read books differently, seeing through the eyes of being a parent as opposed to having been parented. Same with film, art, theater. There's a whole new level of inquiry, new ways to consider love and life and language, and—I'm not quite sure how to explain this, so stick with me for a second—it ups the ante. I have something now that's sacred. I want to be better for him: better mom, better writer, better human being on this planet. I want the world to be better, too. I believe that art has a place in that.

So what am I going to do about it?

What am I going to make?

*

Here are some things I would not have told you:

I'm scared.

Something is wrong.

It's not all in my head.

Please, I'm so scared.

Sometimes I can't get off the floor.

I need help.

I need help.

I need help.

*

I am reading about postpartum depression. I want to understand what happened to me. Books describe hormonal changes, sleep deprivation, and the fear of parenthood—all true for me and, I'd wager, others, too: birth moms and dads and adoptive parents, queer and straight, cis and trans, one-, two-, and three-parent households, and the intersections of class and race and ability. From there, our stories diverge based on individual circumstances and beliefs. To assume that everyone's experience of birth and family and illness looks the same as your experience of birth and family and illness is bat shit at best and, at worst, dangerous as all hell.

Here is the truth: I was lucky. I had a healthy child, health insurance, a supportive partner, and a village.

This is true, too: I was an adjunct college teacher with no paid maternity leave, and Christopher worked ten/twelve hour days. It was early 2008, the first months of the recession. The value of our condo had plummeted and costs were climbing. The winter was awful, gray sky for months. I had a hard time breast-feeding and was terrified to talk about it. The baby was losing weight so we borrowed money to pay for a lactation consultant and eventually figured it out, but still: the mental job I did on myself was the equivalent of a power drill through the shoulder bone, not to mention a then undiagnosed and particularly gnarly thyroid disorder. Inside the mess of my head was a single clear thought: *I want to be saved.*

*

Jesus. How do you write about depression in a way that's not depressing?

*

Lott came over to hold the baby. I locked myself in the bathroom and turned on the water so it sounded like I was showering. Then I sat on the floor in the dark and cried, Mojo's head in my lap. Afterward, I wrapped a towel around my unwashed hair, you know, to look as though I'd washed it. Lott is a good friend. He pretended not to notice.

Jeff came over to hold the baby. I locked myself in the bathroom, turned on the shower, and got under the faucet fully dressed. The water drenched my sweatshirt, my yoga pants and socks. There's commentary here about control: my body was no longer mine, my thoughts were no longer mine, my life was no longer mine, but this ridiculous thing I'm doing now? All me, baby. I sloshed into the living room, my clothes heavy and leaking all over the floor. Jeff looked me up and down and said, "You need to hang out with some mothers."

Dia flew in from Oakland to hold the baby. Her Achilles tendon was shredded from football so she scooted around on a knee walker, up and down the three flights of stairs on her butt. I don't know if she'd talked to Lott or Jeff, or if she just knew, but when I told her I was going to take a shower, she followed me into the bathroom, sitting on the toilet like we used to when we lived together in Humboldt Park. She rocked my son in her arms and talked through the curtain: "He looks just like you! Are you using soap? Look at his little feet! Use the soap. Megan? *Soap.*"

Amanda came over, supposedly to hold the baby, but instead she told me to get dressed. I looked at the window. Snow pounded the glass. "We're not going out in that," I said. "We are," she said, tucking Caleb into his tiny snowsuit. He smiled and reached for her thumbs. "Get ready." I cleaned myself up more or less and we drove to her apartment in Edgewater. Twenty-some people were in the living room, waiting for me to run a writing workshop like I had a thousand times before. But I didn't feel like before. There was no more before. Amanda put her hand on my arm. "It's time," she said. I sat on the floor, my sleeping son in a sling across my chest, and I listened to stories, to love and loss and fear. They got into my head. I went home that night and wrote, just lists in my journal, but still. Amanda made a Google calendar and people signed up to babysit while I ran workshops: week after week, month after month, each sentence another step back to myself. Eventually I felt strong enough to tell a story of my own, this weird little thing about watching another mother on a wireless baby monitor. It turned into something, a contribution, I hoped, to some sort of dialogue.

Postpartum depression—depression, period—is something we don't talk about.

Why?

Shame?

The hell with shame.

Let's get into it, the beauty and the mess.

Today I get e-mails from women who've recently had babies and are there now, in the thick of it. I get e-mails from the men and women who love them, too. I like to think of us all together, crowded into a football field, sitting in the bleachers. We share popcorn. We do the wave. We see that we're not alone.

*

One early Sunday, a few months before we walked away from our mortgage, I let Christopher sleep in and took Caleb to the dog beach. The lake was still, a mirror, and we waded in midcalf and threw tennis balls for Mojo. He'd wait, body rigid, eye on the ball held over my head, ready to spring, and I'd pitch it as hard and far as I could, trying to throw like Mo'ne Davis. When the ball leaves my hand, he's off, a fifty-pound missile cutting over the water, then leaping over the water, then paddling through the water toward his yellow target, closer, closer, success! And back to me, tail wagging, proud as hell, rinse and repeat. It was late August and the water was perfect, a bathtub. Caleb was four years old, splashing around until the sun rose above us and then he called out colors: "Pink! Orange! Yellow! Mommy, look!" I looked. It was incredible, the blue of the sky and the blue of the water with a single stripe of sunlight, a path into the sky.

That's when the singing started.

Behind us, gathered on the beach, people were singing—twenty of them, maybe? Thirty? Black and brown and white:

children, middle-, and third-agers,* all in the sand in their Sunday best and many dressed head to toe in white.

"Mommy, what's happening?" said my son, splashing through the water to grab the hem of my dress. Even so young, having never set foot in a church, he understood the reverence.

"Shhhh," I whispered, leashing the dog. "Watch." One of the men—the reverend, the pastor?—took the hand of an older woman dressed in white, and the two of them walked into the water, their clothes seeping as they went deeper and dragging them slow, the sun climbing higher, the congregation still singing from the shore. I wondered if we should leave, if my little family and I were intruding on this private moment, but no. The reverend, water now up to his chest, turned to wave at the group on the shore and then at my small son. He placed his palms together and nodded toward us.

Welcome.

The sun climbed. The reverend held the woman's hands and they spoke quietly. He asked questions I couldn't hear, and to each, she nodded yes yes yes. Her white, wet dress streamed around them. Then he took her in his arms, dipping her backward and fully underwater.

I held my breath with her.

She came up drenched, wiping her eyes, laughing and crying and god, the joy. She hugged the reverend, then turned to face the beach, throwing both arms into the air as if she'd just scored a touchdown. Her friends cheered, and she splashed

* My friend Bobby works with storytellers in their seventies and eighties in senior living facilities around Chicago. Recently, they told him they don't like the word "senior." Bobby asked them what they'd prefer and they said "third-ager," which I believe comes from the French: teenage, middle age, third age.

back toward them, switching places with another woman walking to the reverend. Behind her splashed a little boy no bigger than my own. He stopped near us, water to his waist. As she and the reverend got ready to duck under, the little boy got ready, too, holding his thumb and forefinger in front of his face. "One, two—" he counted, watching them carefully, and on three he sucked in a breath, plugged his nose, and went down and up as they did.

Everybody cheered from the beach.

The little boy jumped up and down, laughing and dripping.

"Why are we so happy?" asked my son, jumping with him.

"I don't know!" he said.

For the rest of the morning, the two of them played baptism, holding their noses and dunking as the congregation celebrated on the beach. I'd never seen anything like it. I've had glimpses, I think, like when I first heard Sweet Honey in the Rock sing "I Be Your Water"; the Chicago Symphony Orchestra doing *Carmina Burana*; standing on Petrin Hill, looking over late-night Prague; in Alaska on Spruce Island, the forest a carpet of moss; that storefront church in Humboldt Park; giving birth in a snowstorm; the floor of a rowboat on Cedar Lake; my dad's boat in the Gulf of Alaska, all blue and no horizon; Dorothy Allison reading aloud, her voice lifting the ceiling; Joy Harjo on releasing fear; Patti Smith on the moon; and Toni Morrison on learning how to love.

I let the dog off the leash and watched him fly across the sand.

Maybe I was saved.

Maybe we all are.

We Say and Do Kind Things

Sarah and I are drinking. We're at Little Bad Wolf, a cocktail bar on Chicago's North Side. It has fancy tacos and great old-fashioneds. Like, really great.

We've had a lot of them.

A month earlier, her two-year-old daughter, Sophia, was diagnosed with a brain tumor. She couldn't shake the flu, and when Sarah took her to the pediatrician they discovered the flu wasn't the flu.

Snap your fingers.

That's how fast your life can change.

*

Sarah is goddamn fucking sunshine. Weaponized optimism. You'll be all: *I had a shitty day*. And she'll say, *Oh, friend, that's terrible. Put down your things and we'll have a quick dance. Sarah*, you'll say, *we're in a parking lot*. Or: *Sarah, it's raining*. Or: *Sarah, there's no music*, and she will give you a look. You'll drop your stuff. You'll take off your shoes. You'll dance your face off to the sound of water hitting the pavement and you know what? It's glorious.

I hope you have a person like that in your life.

One who reminds you to choose joy.

*

She's the first to say she's blessed: her beautiful little girl; two older boys, patient and sweet; her husband, Scott, in the thick of it with her. They have insurance. They have some work flexibility. They have a village rallying around their family. Within a week of Sophia's diagnosis, our friends had set up a website to organize the kindness: meals, pickups, playdates. It also allows Sarah to send out updates so she doesn't have to talk about it every time she runs into somebody at the grocery store.

"What else can I do?" I asked when she dropped the boys at my house, the first morning of who knows how many rounds of chemo. Her face was locked in a sunny smile: brave for Sophia, for the boys, for Scott. I thought she was a fucking warrior. I thought: *This is what strength looks like.* Not a bodybuilder with the biceps. Not Superman holding up a skyscraper. Not an army of thousands with their guns and their tanks and their bombs—no.

Strength is a mother.

"At some point—" she said through the smile.

"Name it," I said.

"Not today—"

"You say when."

"I will need to get drunk."

*

We met not long after I had my son. My friends were worried about me and thought I needed to talk to other mothers, so Amanda called Sarah. She had every reason to say no. She didn't know me. She had two small kids of her own to take care of, an insane schedule, and who wants to go outside in a Chicago winter?

"Hi!" she said, when I opened the door.

"Hi!" I said, and burst into tears.

I'd never been a crier—"We're Dutch," my father says—but at that point I couldn't stop. At the pediatrician's office, my tiny son slept in his infant car seat while I filled out paperwork including a long list of questions regarding the mother's postpartum health. Number five: *Have you been crying excessively?* Immediately I started crying excessively, there in the lobby with its colored walls and cutesy playroom and happy toddlers toddling around. The receptionist came over with a box of Kleenex, leaned into me, and whispered, "Look at the woman behind you." I turned, and there she was, infant car seat at *her* feet, crying *her* eyes out. "You're totally normal," the receptionist said. "You're all totally normal."

I still carry those words.

A life raft. A lighthouse. The last canteen in a dying desert.

Crying doesn't faze Sarah. "It's lovely to meet you!" she said outside the door, holding a bottle of wine. She wore an impeccable black wool cape and knee-high boots, with long, thick hair to her waist. I stood just inside, holding my crying son—he started whenever I started and vice versa, the two of us caught in a weird ouroboros loop—with snot everywhere and the same yoga pants I'd worn all week, my hair falling out in clumps. "How's about we switch?" she asked, reaching for the baby and handing me the wine.

Life raft, lighthouse, last canteen.

*

If you want to hide out for a night, Little Bad Wolf is your place. It's small and dark, low lights and wooden tables. Sarah and I sat in the corner, mainlined bourbon, and talked about things that weren't cancer.

Then we talked about cancer.

Cancer is here now.

Fuck cancer.

"Sophia finally fell asleep," she said, telling me about the most recent round of chemo. They're in treatment at Lurie Children's Hospital downtown—another blessing, Sarah's quick to mention. The best pediatric cancer facility in the Midwest and it's here. They don't have to leave home like so many families do. "I left her in the room with Scott and took the elevator downstairs. I needed air. Space. Something. I went outside, and—do you know the park across the street?"

We build maps in our heads. I'd been to the ER at Lurie several times, when ear infections made my son think his brain was exploding. He was delivered next door at Prentice Women's Hospital. Our first pediatrician's office is around the corner. I taught down the street at Northwestern's Chicago campus. Activists shut down Michigan Avenue following the cover-up of Laquan McDonald's murder. The Magnificent Mile Lights Festival, now named for a bank. *Alice* at Lookingglass Theatre. Martin Creed's *Mothers* at the Museum of Contemporary Art. I spent hours in what used to be Borders. I buy face cream at Marshall Field's, which I know was bought out by Macy's but to a Chicagoan Marshall Field's is Marshall Field's. Comiskey Park is Comiskey Park. The Sears Tower is the Sears Tower. The *Tribune* is the *Tribune*.

"The park in front of the museum?" I said.

She nodded. "I sat there for a long time. People walked by on their way to work, shopping, playing with their kids, going about their lives as though everything was fine, and I just—I lost it."

Have you held a friend's hand as their heart breaks? This woman took care of me, peeled me off the floor. For her I'd kill a dragon.

Except she doesn't want a dragon. She wants a cure for pediatric cancer.

"I couldn't stop crying," she went on. "All these people were staring at me and none of them—No one—"

"No one asked if you were okay?" I tried not to scream. I'd had a lot to drink. My friend was hurting. Her child was hurting. Words can't fix it, but they are a start: "are" and "you" and "okay." I thought of the times I hadn't said them: people I'd assumed were asleep, passed out, wanted to be left alone, don't draw attention, not your place. What if they were sick? Needed help? Needed shielding from a creepy guy or a loud-ass racist or an online troll or some small act of humanity to make it through the next five minutes? I wanted to show up at that park and yell at everyone walking past my crying friend. I wanted to kick myself for all the times I could have helped but hadn't. I wanted to go to med school and find a cure. I wanted to raise a gazillion dollars for research. I wanted to give Sophia a unicorn. I wanted to hug Sarah but the table was at a weird angle. I wanted a better angle. I wanted a better world. I wanted to be a better person.

"But then," Sarah went on, sweeping hair out of her face. She's a Pantene commercial, that woman. "After I was all cried out, I went back to the hospital and got in the elevator up to Sophia's floor, and a woman got on with me. She was a nurse, I think. She had on scrubs. The doors shut in front of us, and she looked at me, and you know what she said? She said, 'Do you need a hug?'"

I watched the memory on her face.

"It's hard to explain."

Life raft, lighthouse, last canteen.

"I'm afraid," she said finally. "And it was nice to feel something else. Even for only a second."

*

A few months after Sophia began treatment, I spent the day with her and Sarah at the hospital. She wears bunny slippers everywhere. Two pigtails stick out of her head. Sunshine, like her mother. "Here are the fish!" she said, dragging me to a wall-size tank. "Here are the donuts!" she said, waving at the coffee kiosk. "Here are my friends!" she said, introducing me to the nurses and technicians who take care of her every week, weight and measurements and tests and procedures, so many procedures, such a tiny body for so huge a fight.

There are good days and bad days, Sarah says. On good days, they celebrate.

That day was a good day.

There are murals in the elevators at Lurie: happy cities with bright blue skies. Buttons are placed kid high that make traffic sounds, honking and beeping and bicycle bells. Sophia slid across the floor in her bunnies, careful to press each one. She made Sarah press them, too, and me. She made us press ones on opposite walls at the same time, and then we'd switch, the three of us dancing around the elevator, honking and laughing and you know what? It was glorious.

The doors opened then and a woman got on alone. She wore sweatpants. Her eyes were red. Her hands shook. Her breath shook. She smiled at Sophia, nodded to Sarah and me, and then stared ahead frozen as the elevator counted down.

The air was heavy. Grief near tangible.

"Excuse me," Sarah said, gently touching her arm. "Do you need a hug?"

And for the rest of the ride and a half hour downstairs in the lobby, I watched two-year-old Sophia watch her mother be kind.

*

My mother is an elementary school teacher. There's one rule in her classroom: WE SAY AND DO KIND THINGS. It's spelled out in enormous construction paper letters hung across the blackboard, starting at one wall and ending at the other. She and her kids talk about it on the first day. What does it mean to be kind? It's the beginning of an ongoing conversation that continues throughout the school year and hopefully the rest of their lives.

In our house, it was a sort of mission statement, the lens she held up to show me the world. At first it was simple. Say please. If you hurt someone, apologize. Ask how you can make it better. Share your stuff. Don't take things that belong to someone else like when my cousin Aaron swiped Evil-Lyn, my favorite Masters of the Universe action figure and why was she second in command to Skeletor when clearly she was more powerful? We get older and kindness, along with everything else, gets complicated. The guy in the bar who grabbed me and would only let go if I said please. Apologies saying *I'm sorry you were offended* as opposed to *I'm sorry for what I did* and *Here's how I'll make it right*. Sometimes I wonder how the world would be different if we shared resources and opportunities the way our kindergarten teachers made us share crayons. I remember coming home furious from a high school history class on colonization, a screaming example of taking what belongs to someone else. "We took *lives*," I told my mother. "We took *land*. We took *culture*." I was a small-town, sheltered white kid. I was all guilt and rage. I stood in my kitchen and yelled the understatement of the century: "IT ISN'T FUCKING KIND."

"It's not," my mother said. "How can we make it right?"

Sometimes kindness means making it right.

Sometimes kindness means showing up.

Sometimes it's trying.

We have to try.

*

The day after the hospital with Sarah and Sophia, I kept my son home from school. Life moves so fast—how was he already seven?—and I wanted to hit Pause. We ran on the beach. We hung out at the bookstore. We had lunch at Costello's, our favorite sandwich shop, and I told him a story about our family.

"During the Great Depression—" I started.

We stopped to discuss the Great Depression.

"My great-grandparents—"

We stopped to discuss genealogy.

—opened their farmhouse in Holland, Michigan, to families in need of a place to sleep, a hostel of sorts through the church. My great-aunt Jennie took me into the basement to show me the cots, the marks on the door where kids had carved their names. "Hebrews 13:2," she told me. I was twenty years old, at a complicated place with religion, but being with Aunt Jennie made me think about the church as a place of kindness as opposed to fear. "Be not forgetful to entertain strangers," she recited, and in her voice I heard the poetry, "for thereby some have entertained angels unawares."

We stopped to discuss angels.

"They take care of us, right?" he asked, mouth full of chips.

"That's what I believe."

He considered, chewing. "And an angel can be anyone?"

A nurse in an elevator. A receptionist in a pediatrician's office. A mother in your hallway. Her daughter reminding you to dance. "Yes," I said.

He looked around the restaurant: the woman at the next table, the guy behind the counter, all the people in line. "Anyone?" he said. "Even you?"

"Anyone," I said. "Even me."

After lunch we headed downtown, Lake Shore Drive to Michigan Avenue. He watched the buildings sprout around us and asked, "Are we going to see Sophia?" thinking of the times we'd visited her at Lurie.

"Nope," I said.

He thought for a second. "Are we going to see David Bowie?"

That was the map in his head: visiting his buddy at the hospital and a retrospective on Bowie at the MCA.

We drove into the parking structure, around and around until we found a spot, then down one elevator and up another. The walk was muscle memory, months lugging the infant car seat and scared out of my mind. He wasn't gaining, wasn't latching, wasn't sleeping, and I'd sit in the lobby and sob. It still looked the same: colored walls, cutesy playroom, happy toddlers, and, behind the desk, the receptionist who brought me Kleenex. I knew she'd be there. I'd looked at the staff page online and found her picture, then called to ask if she was working.

"Why are we here?" asked my son.

"I forgot to do something," I told him.

She looked up, smiled, asked, "Can I help you?" And I gripped the desk with both fists and took a deep breath so I wouldn't cry and said in a rushing run-on sentence that I was sure she wouldn't remember but seven years before I had severe postpartum depression and one day at an appointment I started crying in the lobby and she came over and told me I was normal and that seemingly small gesture was a life raft in ways I'd only begun to articulate and I wanted her to know how grateful I was and how grateful my family was and I hoped I wasn't freaking her out and also she's an angel.

I exhaled.

She reached out and put a hand on mine. "How are you?" she said.

"I'm good," I said. Telling her made it true.

We were still for a moment, holding hands. And then, very purposely, she turned her head toward the lobby. I followed her eyes and there—the God's honest truth—sat a woman. There was an infant car seat at her feet. She was crying.

I looked back at the receptionist. I wondered how often this happened, how many of us she had saved. We had a brief conversation that did not involve speaking. "What's your name?" she asked my son, and as they chatted, I picked up the box of Kleenex from the desk and went into the lobby.

*

I walked into the Halloween party at our kids' elementary school, running late from work. I was tired and my feet hurt and the bag of candy I carried was splitting at the base. Hundreds of costumed children ran around screaming. They pulled streamers off the ceiling. Sugar hit the bloodstream. I said hello to Sarah's boys—a Roman gladiator, a purple hawk—and looked around for Boba Fett. "Hi, Mom!" he said, appearing at my elbow. "Bye, Mom!" he said, and back into the crowd.

Sarah was onstage, leading the dance game where you freeze when the music stops. She was wearing an orange-and-black tutu and orange-and-black-striped tights. Weird glitter antennae jutted out of her head and her crazy-gorgeous hair was cropped at the chin, cut off the spring before at a fundraising event for St. Baldrick's. Her goal was $500; in the end, her hair brought $40,000.

Sometimes kindness means money.

I felt a tug on my skirt and looked down. It was Sophia, dressed as Little Red Riding Hood with a long cape and an enormous wicker basket. It was full of lunch meat and cheese cubes and fruit and popcorn and cookies and candy mixed together. Obviously, she'd gone on a solo run to the snack table. "Here is the salami!" she said, holding up her basket.

I put down the candy and squatted so we were eye level. "How are you, Sophia?" I asked. She was three years old, in treatment for just under a year.

There are good days and bad days.

On good days, you celebrate.

"Come on," she said, reaching for my hand. "We have to dance!"

I started to tell her no—tired, feet hurt, broken candy bag—but she gave me a look.

I knew that look.

So I kicked off my shoes, climbed onstage, and danced my face off with my friend in her crazy tutu and an angel of a little girl and you know what? It was glorious.

The Wrong Way to Save Your Life

I was getting ready for bed when I heard sirens. Not a big a deal in the city, right? We hear them all the time. I brushed my teeth—they got louder. I peed—louder. I took off my pants—louder, closer, the dog was going bat shit. And when I came out of the bathroom I saw red lights splicing through the blinds on my front windows, red stripes running down the living room walls. "What the—" said my husband, getting up from the couch, and that's when we heard pounding on the door.

*

At some point in my early twenties, I made a list, scribbled on the back of a cocktail napkin, stuck in an old journal, and then forgotten. Until now, when my life caught up to its question.

The napkin is crumpled, my long-ago handwriting shaky and near illegible. For sure I was tipsy if not totally lit, hanging with friends at a bar and killing time with hypothetical games: Truth or Dare. Two Truths and a Lie. The Proust Questionnaire. *What five things would you bring to a desert island? What five things would you take to the moon? What five things would you need for the zombie apocalypse?* Or, in the case of this particular list: *What five things would you grab in a fire?* I vaguely remember the drunken setup: "You wake up choking, the room's full of smoke, flames everywhere, you've got two minutes to get out before you roast alive. So what do you grab? What is *that* important?"

Here's what I wrote on the napkin:

1. Journals.

While I applaud my younger self for wanting to save her memories, this choice is totally illogical. Back then I used big green hardbound sketchbooks, stored under my futon in tomato boxes from Jewel-Osco. No way I could have carried them all out in time.

Later, sick of lugging them from apartment to apartment, I stashed them in my mom's attic in Michigan. "Aren't you afraid she'll read them?" a friend asked, shocked I'd let one of my parents so close to my secrets. Honestly, there's nothing she doesn't already know. I tell my mother everything. I always have. I called her sobbing from a pay phone in Italy after walking out of that gynecologist's office. When I called her from Prague and told her I was in love, she got on a plane the next day to come meet him. She Facetimes with her grandson every morning so I can sleep for an extra ten minutes. In the chaos following the school shooting in my hometown—no cell phones, landlines backed up, no one able to get a hold of each other, "the victim was a local school administrator"—she got the person who ran the front desk at my dorm in Boston to put her through to an administrator who passed her up the chain all the way to the university president, making him promise to find me right then and tell me that my father was alive.

After years of reading personal essays by students and storytellers and the complex, often painful relationships they have with their own mothers, I fully understand the gift of having a woman I trust, who supports me unconditionally and loves me fiercely.

2. A photo of my parents.

My uncle Mike sent it to me. He found it in a box that belonged to my grandmother. I am tiny, in a diaper, trying to walk to what I think is a tree. My dad is wearing cool sneakers. My mom is a total fox. They're laughing.

Look: they're both happy now.

But it's nice to see that they were happy then.

3. My first edition Anaïs Nin.

I did not own a first edition of anything, so the inclusion of Nin on the napkin list is pure pretention on the part of twenty-year-old me; however, I must have mentioned this material desire to my husband at some point because he gave me a rare signed interview with her as an anniversary gift. It lives in a safe-deposit box at the bank and is therefore not at risk for fire damage, real or hypothetical.

Now, if you'll indulge what could be construed as pretension from forty-year-old me: occasionally I go to the bank and take the interview out of its box. I pet the fragile, letterpress pages and reread her words. They've been helpful as I consider what it means to be a woman telling her own stories: "Our culture has put a taboo on what they call introspection and the growth of the self. Any attention you gave your own world was sinful. And I think what I get mostly in the letters people write to me is that this is a study of growth."

4. Anything that said Marc Jacobs.

This is just ridiculous. I didn't own anything made by Marc Jacobs. I couldn't afford Marc Jacobs. I couldn't fit into Marc Jacobs.

Dear Marc Jacobs: I like your use of mismatched patterns and I have a big, lovely ass. Please make me some pants.

5. Birth control.

Good girl.

*

We live on the second floor of a three-flat rental in Rogers Park. It's your typical Chicago layout: in the front, living room and bedroom number one; in the middle, dining room, bathroom, and bedroom number two; in the back, kitchen and bedroom number three. We'd moved in a year earlier, after bailing on the condo we bought just before the market crashed. It happened so fast: one day we had a home and then—snap your fingers—it was worthless, underwater they call it, which is a poetic way of saying that you're drowning. We couldn't sell. We'd just had a baby. I was sick. My husband took a second job. I got better and took a third, then a fourth, sixty-hour weeks of hustling, passing our son between us like a football. It was this wild whiplash between gratitude—we had the jobs, the home, each other—and exhaustion.

One day I woke up and five years had passed.

At some point you have to ask yourself how you want to live.

I typed the word "beach" into Craigslist and an apartment came up in Rogers Park: typical Chicago layout. Steps from the lakefront. Working fireplace. Rent was half our monthly mortgage. My husband and I talked about what would happen, to our credit, to our future. And a week later, I sat on my new front porch, listening to Lake Michigan waves against the sand. My then five-year-old son sat on my lap. I felt him breathing: Slow. Easy. Calm.

"It's nice to be home," he said.

It was that first deep drink of air after coming up from underwater.

So imagine how it felt when my upstairs neighbor pounded on our door and told us the building was on fire.

*

In my late twenties, I sold my stuff and moved to Prague. I wanted to make art. I wanted to reset. I wanted to be light. I took boxes to Myopic Books and posted the big stuff on Craigslist. My journals were already at my mom's; everything else had to fit in a backpack.

1. Passport.

I arrived in Prague during the 2004 presidential election between George Bush and John Kerry, a year into the United States–led invasion of Iraq, which the majority of the Czech people vehemently opposed. Everyone I met asked how my country could do such a thing, and "I don't know" was not a good enough answer. Neither was "I don't agree with my government" or "I don't have a say in my government." This was the home of the Velvet Revolution, a massive antigovernment protest that began with a student march, grew in theaters, and culminated with 750,000 people in the streets as the Communist Party stepped down. To stand in Wenceslas Square and say my voice didn't matter was ridiculous. I was an educator, an artist, a citizen.

My voice is more than my vote.

Here's the truth: I hadn't yet done the work of examining my complicity in systems that were hurting people, at home and abroad. I didn't understand the power in my passport. I didn't know that I could be furious with my country and, at the same time, cherish it.

The Klementinum Library is one of the most beautiful in the world: baroque architecture, frescoes on the ceiling, opened in 1722. It's all rare art and priceless books, not the place to camp all day with my coffee and my questions. Instead, I went to the Internet café down the street from my flat, ordered a pivo, and googled: how is America viewed abroad?

I was there for a really long time.

I'm there still.

2. Laptop.

When I was a kid, I had a mild obsession with *Inspector Gadget*, a television cartoon about a wacky cyborg detective. He had a niece named Penny who I totally wanted to be, in part because she was a genius but mostly because she had a book *that was actually a computer.*

I wonder what eight-year-old me would say about twenty-eight-year-old me, sitting in a kavarna every morning and writing so nonchalantly on such a miraculous device.

I wonder what my now eight-year-old son will think of his twenty-eight-year-old self, walking on virtual Mars or SWEEPS-11 with a microchip implanted in his eyeball.

Something like that, right?

3. Franz Kafka's *Diaries: 1910–1923.*

I was there to teach a class on Kafka, not as literary theory but writing for writers, the process and the mess. I'd studied his work for years. He fascinated me. He terrified me. Reading his journals pushed me into my own, so much wonder and fear in my head and my heart and God, what a relief to get it out of me, to read what I think and see what I say.

You can't fix it if you can't see it.

One part in particular in *Diaries* that I keep coming back to: a section near the beginning, from the summer of 1910: "When I think about it," he writes, "I must say that my education has done me great harm in some respects." He talks about this for a paragraph, then stops and begins again: "I must say that my education has done me great harm in some respects," and on for another paragraph, digging deeper, until again he stops and starts again: "Often I think it over and then I always have to say that my education has done me great harm in some ways," and on again, still deeper. This repeats six times, each section coming at the idea from a different place and arriving at different ideas.

I use (read: steal) this structure when I'm trying to understand what feels impossible. My fear that the depression will come back, that I will wake up in the morning and not be able to get up off the floor. My fear that something will happen to my child and my heart will split open and a tidal wave of blood will pour out of my body and flood the streets. My fear that members of my family, blood and chosen, will be harmed because of who they love. My fear that members of my family will be harmed because of the color of their skin. My fear of being a person who is concerned only for members of her own family and not all families.

4. Knife.

It's a Bucklite MAX with a guthook.
A present from my dad.

5. Birth Control.

Good girl.

*

My upstairs neighbor stood in the hall, red lights strobing, sirens screaming, dog whimpering, not noticing or not caring that both my husband and I were in our underwear. They'd had a fire in the fireplace, he told us. It died out like always, they went to bed, were about to fall asleep, but something made him get up again, a feeling I have all the time. You're lying there, so tired, so comfy, almost asleep and then: *Shit, did I forget to lock the doors?*

If you'll permit me here a little PSA: I don't care how snuggly the sheets, I don't care how warm the bed, I don't care how many times you've checked: if your gut says get up then GET THE FUCK UP because when my neighbor went into his living room, he saw the wall behind his fireplace burst into flames.

*

In my early thirties, we bought the condo. I remember being shocked at how quickly we became adults. We had passwords. Documents in triplicate. Our signatures were required here and here and here and our initials here. I found myself researching safe-deposit boxes, what should be stored at the bank and what should be at home, immediately accessible and sealed in waterproof bags inside a portable titanium biometric security safe with a fingerprint reader kept in the refrigerator for added precaution.

I may have been watching a little too much *Alias*.

I may have been spending a little too much time on the Internet.

I may have already been pregnant, carrying sleeves of saltine crackers in my purse and worried about the world. I wanted to save everything, all of us, to summon the superhuman strength mothers get when they lift trucks off of their children; instead, I filled out paperwork. I got things notarized. I tried to be prepared.

1. Birth certificates:

—in case we died.

2. Power of attorney:

—in case we died.

3. Life insurance:

—in case we died.

4. Deed to the condo:

—in case we died trying to sell it which, believe you me, almost happened a couple of times.

5. Letter to my little boy, telling him how fiercely he is loved.

*

Our upstairs neighbor ran to warn our downstairs neighbors, leaving our front door open. From the hallway, I smelled smoke. The sirens had multiplied—three fire trucks outside now, maybe four. I turned toward the front of the apartment: our books and art and photos lit red and flashing, the closet with the lockbox. Then I turned to the back, where my now six-year-old son was fast asleep in his bedroom off the kitchen. There were twenty steps between him and me. Twenty steps and we'd be out the back door, down the back stairs, into the car, and out of there.

"I'll get him," I said to my husband.

"Great," he said, already moving, dog at his heels. "I'll grab some stuff and meet you in the car."

I remember thinking: *Stuff*? What stuff? How do you decide, the clock ticking? I vaguely remembered the list I'd made in my twenties. Back then, objects were sacred, not people. Back then, I hadn't experienced loss. Back then, love meant something different entirely.

The question isn't: what would you grab in a fire?

It's: what has meaning in our lives?

*

1. Kid.

Ask most mothers what we'd grab in a fire and the answer is easy: kid(s). But it doesn't mean our being mothers is the only thing that matters. It took me a while to figure that out. So much of what's sold to women is that motherhood is our purpose as opposed to our choice, that we have to have children and put the other parts of our selves at best second, and at worst away for good. I'm here to join the chorus of fuck that noise. If you want to be a mom, be a mom. Be a mom and a working artist and whatever the hell else you want and yes, you will make work after the baby comes and yes, it will be hard and yes, you will be tired but more than that, a thousand times more, it will be amazing and life changing in ways I'm only beginning to understand. And if you don't want to have a kid, if you choose not to go that way, then I'm standing behind you, too, cheering my face off because what has meaning in this life is living it full and true.

*

On the way to my son's room, I grabbed a pair of pants, the same ones I'd taken off in the bathroom mere moments before. I put a toothbrush in one pocket and moisturizer in the other. Then I ran into my bedroom and, for reasons I still don't fully understand, grabbed the knife from my bedside table, the same one my dad had given me years ago, and slid it under my waistband. How long did that all take? Thirty seconds? A minute? But during that time my husband, still in his underpants, ran to the back of the house carrying our laptops, backup hard drive, and necessary cords. It's fair to say that he values technology. It's expensive, of course, but it's more than that: In those laptops is the website he built, the one that supports our family. They also hold the book I was working on, essays about fear. It didn't occur to me to grab my own book.

What occurred to me was: Pants. Toiletries. Knife.

I've examined these objects obsessively, trying to figure out why they seemed so necessary in the moment, a weird reverse of my long-ago game. Instead of: *what would you grab in a fire?* It's: *why the hell did you grab* THAT? Some of my reasoning is logical: it was January, cold as hell, hence pants. Some of it is ridiculous: we'd been watching *The Walking Dead* so, instead of grabbing something to help me in a fire, I grabbed a knife for the zombie apocalypse. Some of it, I'm now realizing, is an attempt to process loss: my writing isn't in the hard drive, the Dropbox, the journals. It's a practice, a process, like my friend Pete taught me years ago with the tubes and the sketches and the mess. You can't grab it as you run out the door, can't hold it in your hands any more than you can hold your own heart.

There were twenty steps from him to me, an invisible line between us.

I stayed on that line.

The mind's got nothing on the gut.

*

My kid sleeps through everything. After years of falling asleep across the street from a rock club, you can run a marching band through his room and he won't budge. Sirens were still blaring out front—were there five trucks now? Six? Feet pounded up the hall stairs, god-awful noise coming from above our living room but he would not wake up. I slung him, still sleeping, over my shoulder, ran down the back stairs, and into the snow-covered alley. I wore pants, but no shoes. My feet were so cold they were on fire. I slid into the front seat with him on my lap, next to my husband who was already behind the wheel. The heat was blasting. The dog cowered on the floor. In front of us, our building loomed three stories high, enormous and eerie and backlit red and flashing.

It happened so fast: One day we had a home and then—

Snap your fingers.

I imagined our charred living room, piles of black dust, furniture drenched from fire hoses. In my head I was already fighting with insurance companies, living out of a hotel, calling our parents for help. We could call our parents for help. We had help. What if we didn't have help? What if we hadn't made it out? What if I hadn't got to my son in time, what if I'd tripped running barefoot down the stairs, what if the fire caught up with us? I can still feel the heat, the fear—can't move, can't cry, breath locked. I talk about setting walls on fire and worry that I jinxed us. I talk about climbing through fog and feel it surrounding me, thick like soup. Here is my heart, laid out in the open, and when I look, *really* look, I don't always like what I see.

You have to see it if you want to fix it.

"Where should we go?" my husband asked.

I listed names, and with each remembered how lucky we were. Our friends would take us in. Our kid wouldn't remember a thing. Our upstairs neighbor had gotten up to check his locks. Our downstairs neighbor, we'd later learn, was a thirty-year veteran of the Chicago Fire Department. He'd run upstairs in his pajamas and attacked the third floor with a crowbar, ripping clear a pathway for the hoses and extinguishers and chemicals that wiped the flames away before they spread. This was one of a thousand nights that could've gone one way, but it went another.

Not long after, we'd get the call that it was safe, but just then we drove down the alley, leaving yet another home behind.

I wrapped my body around my son's, feeling him breathe: Slow. Easy. Calm.

I can still feel it.

A memory not in my head but my bones.

Tools or Weapons, Depending on Your Translation

Angels in America, Tony Kushner
Asking for It, Kate Harding
"The Aquarium," Aleksandar Hemon
Bad Feminist, Roxane Gay
Beloved, Toni Morrison
bitches gotta eat, Samantha Irby
Bloom, Anna Schuleit
The Bluest Eye, Toni Morrison
Bluebirds Used to Croon in the Choir, Joe Meno
The BreakBeat Poets, edited by Kevin Coval, Quraysh Ali Lansana, and Nate Marshall
Buffy the Vampire Slayer
"Building the Man I Am," Thomas Page McBee
"The Case for Reparations," Ta-Nehisi Coates
The Chronology of Water, Lidia Yuknavitch
Citizen, Claudia Rankine
The City of Lost Children, Jean-Pierre Jeunet
Colossal
Crazy Horse's Girlfriend, Erika T. Wurth
Crown Fountain, Jaume Plensa

"Dancing Outside Yourself," Khanisha Foster
"The Danger of a Single Story," Chimamanda Adichie
Dark Sparkler, Amber Tamblyn
"Darkness, then Light," Deb R. Lewis
"Dear Straight People," Denice Frohman
"Dear White Moms," Keesha Beckford
Dept. of Speculation, Jenny Offill
Don't Kiss Me, Lindsay Hunter
"Done," Frazey Ford
"Door to Door," Scotty Karate
Drawing Blood, Molly Crabapple
The Empathy Exams, Leslie Jamison
Eva Luna, Isabel Allende
"Every Single Night," Fiona Apple
Excavation, Wendy C. Ortiz
The Faraway Nearby, Rebecca Solnit
The Fifth Element, Luc Besson
The Fire Next Time, James Baldwin
Firefly
The First Collection of Criticism by a Living Female Rock Critic, Jessica Hopper
"The Fourth State of Matter," Jo Ann Beard
Free Street Theater
From Doom to Boom, Collin van der Sluijs
*The F***ing Epic Twitter Quest of @MayorEmanuel*, Dan Sinker
"Good Bones," Maggie Smith
"Groundhog Day," Corin Tucker
"The Guns of My Girlhood," Ann Patchett
Hamilton, Lin-Manuel Miranda
"Heartbeats," José Gonzalez

HEAVN, Jamila Woods
Her Story
Him, Me, Muhammad Ali, Randa Jarrar
Hope Is the Thing with Feathers, Diana Sudyka
How to Slowly Kill Yourself and Others in America, Kiese Laymon
"Human Behavior," Björk
I Am Still Fighting with My Big and Small Fears. But Lately I Seem to Be Winning, Jasmin Siddiqui, (Hera of Herakut)
"I Like Giants," Kimya Dawson
"I Wish I Was the Moon," Neko Case
"If I Had a Boat," Lyle Lovett
"Immaculate Heart College Art Department Rules," Sister Corita Kent
The Impossible Will Take a Little While, edited by Paul Rogat Loeb
In This Land, Sweet Honey in the Rock
"Jazz Music," Bobby Biedrzycki
"Kubuku Rides Again (This is it)," Larry Brown
"Landslide," Stevie Nicks
The Last Illusion, Porochista Khakpour
Lean With It, Paul Octavious
Lemonade, Beyoncé
"Let's Go Crazy," Prince
Letters to a Young Artist, Anna Deavere Smith
Letters to a Young Poet, Rainer Maria Rilke
"Living Like Weasels," Annie Dillard
"Just Looking for Trouble," Hafiz
*Love Medic*ine, Louise Erdrich
Louder Than a Bomb
Mad Max: Fury Road, George Miller

"Mad Rush," Philip Glass
"Madonnas and Whores: On Mothers Writing About Sex," Gina Frangello
Magic for Beginners, Kelly Link
The Matrix, Lana Wachowski and Lilly Wachowski
Me, My Mom and Sharmila, Fawzia Mirza
Measuring the Universe, Roman Ondák
Meet Me in the Moon Room, Ray Vukcevich
The Miseducation of Lauryn Hill, Lauryn Hill
My Only Wife, Jac Jemc
The New Jim Crow, Michelle Alexander
No Hay Mal, Lily Be
"No Place for Self-Pity, No Room for Fear," Toni Morrison
Notes from No Man's Land, Eula Biss
The Obliteration Room, Yayoi Kusama
"OCD," Neil Hilborn
"On Seeing the 100% Perfect Girl One Beautiful April Morning," Haruki Murakami
"One Source," Khuli Chana
Orphan Black
Pedagogy of the Oppressed, Paulo Freire
"Pluto Shits on the Universe," Fatimah Asghar
"I Give It Back: A Poem to Get Rid of Fear," Joy Harjo
The Princess and the Warrior, Tom Tykwer
"Rape Fantasies," Margaret Atwood
Raven Girl, Audrey Niffenegger
Reading Club, Cinta Vidal
Reading Lolita in Tehran, Azar Nafisi
"a remix for remembrance," Kristiana Rae Colon
"River of Names," Dorothy Allison
Rookie

Safety Fifth, Mucca Pazza
salt., Nayyirah Waheed
Same Sex Symbol, Cameron Esposito
"Save Me," Irma
Scandal
The Scared Is Scared, Bianca Giaever
"Searching for Eve," Meredith Talusan
"Self-Portrait of the Artist as an Ungrateful Black Writer," Saeed Jones
SEXomedy, Melissa DuPrey
Sister Outsider, Audre Lorde
Slouching Towards Bethlehem, Joan Didion
Smart Girls at the Party
"Soy Yo," Bomba Estéreo
Staying Alive: Real Poems for Unreal Times, edited by Neil Astley
Stories from the City, Stories from the Sea, PJ Harvey
"Summer in the City," Regina Spektor
Syllabus, Lynda Barry
Symphony City, Amy Martin
Teaching to Transgress, bell hooks
Teen Vogue
The Telling, Zoe Zolbrod
"Thanksgiving Poem, 2012," Coya Paz
Their Eyes Were Watching God, Zora Neale Hurston
"They Pretend to Be Us While Pretending We Don't Exist," Jenny Zhang
Tiny Beautiful Things, Cheryl Strayed
Tonight, We Fuck the Trailer Park Out of Each Other, C. Russell Price
"*Toward a Pathology of the Possessed*," Esme Weijun Wang

"True Colors," Cyndi Lauper
"A Very Old Man with Enormous Wings," Gabriel García Márquez
Vessel, Parneshia Jones
Walden, Henry David Thoreau
The Watch, Hebru Brantley
"We're Not Good Enough to Not Practice," Kiese Laymon
"We Belong," Pat Benatar
"We Didn't," Stuart Dybek
Weather Systems, Andrew Bird
"What Adults Can Learn from Kids," Adora Svitak
When the Messenger Is Hot, Elizabeth Crane
Witness!, Jay Ryan
Work Hard and Be Nice to People, Anthony Burrill
A Wrinkle in Time, Madeleine L'Engle
"The Yellow Wallpaper," Charlotte Perkins Gilman
"You Are the Everything," R.E.M.
You Can Fly Higher, Joseph "Sentrock" Perez
You're So Talented, Samantha Bailey
Zen in the Art of Writing, Ray Bradbury

Acknowledgments

Emily Griffin and Meredith Kaffel Simonoff make me want to climb higher. Every day they challenge me to see this world as something bigger than myself, and I am beyond grateful that they were on the other side of these pages. My love to Cal Morgan and Maya Ziv for their fierce support, at the beginning and still, and to the dream team at Harper Perennial, especially Amanda Pelletier for believing in me, Joanne O'Neill for my heart, and Paula Cooper for the epic conversations in the margins.

Thank you to my teachers. This book is for you.

Thank you to the young writers I've been lucky enough to work with. Your intelligence, vision, and discipline are contagious. Look out, future.

Thank you to Rachel Jamison Webster, Eula Biss, and the Department of English at Northwestern University for bringing me on board; the Ragdale Foundation for the gift of time and space; and the Bongo Room who took care of me while I figured out what the hell I was doing.

Thank you to Soo La Kim, Maggie Ritter, David Noffs, Brian Block, and Ashley Kennedy Makdad. I am so proud to have been part of your team. Thanks, too, to the hundreds of educators we learned from over the years, especially Jennifer Peepas, who so graciously allowed me to write about her work.

Thank you to 2nd Story, *The Paper Machete*, and the Chicago literary community for giving me a home among brilliant artists. You light the best of fires under my ass, and I couldn't be more grateful.

Thank you to the editors who reached down from the clouds in the moments I most needed you: Cheryl Strayed, Roxane Gay, Gina Frangello, Clay Risen, Emily Schultz, Jennifer Niesslein, Lauryn Allison, and Zoe Zolbrod.

Thank you to the many wonderful people who touched this book in profound and mysterious ways: Sarah and Sophia Zematis, Jess and Gus Tschirki, Bobby Biedrzycki, Amy Martin, Amanda Delheimer Dimond, Aaron Stielstra, Ryan Meher, Khanisha Foster, Deb Lewis, Adam Belcuore, Samantha Irby, Elizabeth Crane, Jennifer Pastiloff, Nicole Piasecki, Amy Danzer, Anna March, Molly Each, Kristin Lewis, Elsie Kitchen, Craig Jobson and Judith Grubner, the Zemans, the Sudyka-Ryans, the Kretas whom I love so very much, and my secret online communities of women who make me laugh when I want to stick a fork in my eye.

Above all else: thank you to my family.

Randy Albers, Jeff Oaks, Lott Hill, and Dia Penning hold me together every day. Without them I would have drifted untethered into space like twenty years ago.

My mom read to me and my dad told me stories, and they both love me like mad and the feeling is so totally mutual.

Christopher and Caleb Jobson. Here is my heart.

About the Author

Megan Stielstra is the author of *Once I Was Cool* and *Everyone Remain Calm*. Her work has appeared in *The Best American Essays*, the *New York Times*, *Chicago Tribune*, *Guernica*, *BuzzFeed*, *The Rumpus*, and elsewhere. A longtime company member with 2nd Story, she performs regularly in Chicago and has told stories for National Public Radio, Radio National Australia, the Museum of Contemporary Art, Goodman Theatre, the Neo-Futurarium, and *The Paper Machete* live news magazine at The Green Mill. She teaches creative nonfiction at Northwestern University.